DIY MEMS

Deborah Munro

DIY MEMS

Fabricating Microelectromechanical Systems
in Open Use Labs

 Springer

Deborah Munro
College of Engineering
University of Canterbury
Christchurch, New Zealand

ISBN 978-3-030-33075-0   ISBN 978-3-030-33073-6   (eBook)
https://doi.org/10.1007/978-3-030-33073-6

This Springer imprint is published by the registered company Springer Nature Switzerland AG
The registered company address is: Gewerbestrasse 11, 6330 Cham, Switzerland

This book is dedicated to Dr. Munish Gupta, orthopedic surgeon, for taking me on as a graduate student, believing in me, and allowing me to work on this fascinating research project with him. Without his initial project idea, none of my journey into MEMS would have happened. I would also like to thank Dr. Norman Tien, electrical engineering professor, for teaching me everything I needed to know about MEMS and providing me ongoing guidance on my research over the years. Finally, my sincerest gratitude goes to Dr. Fadi Fathallah, my faculty and research advisor, for being my staunchest supporter throughout my graduate studies. All three of you made this book possible, and I can never thank you enough.

Acknowledgments

I would like to acknowledge Dr. Michael Khbeis, Dr. Andrew Lingley, and Dr. Karl Bohringer of the University of Washington's Nanofabrication Facility (WNF) for collaborating with me on my MEMS project development and training me on how to use their equipment. In particular, I would like to thank Dr. Andrew Lingley for his extensive work in developing the users' manual and training materials at WNF. Without his contributions, gathering the detailed information about lab protocols would have been much more difficult. I would also like to acknowledge Dr. Jiangdong "JD" Deng of the Harvard University Center for Nanoscale Systems (CNS) for consenting, along with Dr. Michael Khbeis, to be interviewed for this book.

Contents

Chapter 1
Introduction

The future of orthopedic medical devices is small—microscopically small. Microelectromechanical systems, referred to as MEMS devices, are already used in a host of electronic devices, doing tasks that were previously only conceived of in science fiction. Sensors the size of a flea can now measure strain or temperature or resistivity. They can also measure acceleration, frequency, and electrical impulses. MEMS devices are used in microscopic gears for tuning hearing aids, capacitive rings embedded in contact lenses to measure glucose levels, and microfluidic pumps to deliver insulin to patients.

However, very little of this technology has infiltrated the medical device market, especially large markets like orthopedic implants, where designs have not changed significantly in the past 40 years. Orthopedic implant manufacturers tend to be conservative, using established, trusted techniques, but recently some companies have successfully incorporated sensors into their devices. For example, there is a total knee arthroplasty soft tissue balancing instrument that measures the applied load and optimizes the selection of the polyethylene tibial component's thickness.

Adding functionality and capability makes a lot of sense in medical devices, as we are all trying to provide our customers more feedback and diagnostic information, and we definitely try to continuously improve patient outcomes. Thus, it will only be a short time before many new products are introduced with embedded electronics and sensor. To be a part of this paradigm shift, now is the time to learn how to incorporate MEMS into your next big design idea.

This book covers what MEMS are, how a device or sensor is fabricated, and where public laboratories are located. It also covers various ways of collaborating with a public laboratory in detail and includes what the main types of equipment are in these facilities. A thorough discussion of intellectual property and privacy issues and how to navigate other issues that may arise is also covered.

Ideas for where and how MEMS could be used in your medical device are limited only by your imagination. For orthopedics, I can envision instruments and implants that provide feedback to the surgeon or the patient, devices that measure the strength of bone or sense the location of an artery or nerve, or cameras that guide

D. Munro, *DIY MEMS*, https://doi.org/10.1007/978-3-030-33073-6_1

the placement of screws. The key is to understand MEMS technology and its possibilities so that you can then see where you could use it in your design.

What Is MEMS?

MEMS is an etching-based machining technique that is done on silicon wafers. Sometimes, layers of material are first deposited, but etching away material is typically required. This allows easy fabrication of features with micron-level dimensions. Do you need a strain gauge that is only a fraction of a millimeter in size? MEMS is the answer; in fact, these are called semiconductor or piezoresistive strain gauges, and you can buy them off the shelf. They are adhesively mounted and attached with wire leads the same way as copper foil strain gauges, except that a microscope and a very steady hand are required. In spite of being tiny, they are actually at least *one hundred times* more sensitive than copper foil strain gauges, because they respond to changes in the crystalline structure. The disadvantage is that they produce a nonlinear response, so they have to be used with a correlation curve or equation to interpret.

According to MEMSnet (https://www.memsnet.org/about/what-is.html), microelectromechanical system, or MEMS, is a technology that in its most general form can be defined as miniaturized mechanical and electromechanical elements that are made using the techniques of microfabrication. Dimensions of MEMS devices can vary from well below one micron on the lower end all the way to several millimeters. MEMS devices can vary from relatively simple structures having no moving elements to extremely complex electromechanical systems with multiple moving elements under the control of integrated microelectronics.

While the functional components of MEMS are miniaturized structures, sensors, actuators, and microelectronics, I believe the most interesting components are the microsensors and microactuators, also known as "transducers," which are defined as devices that convert energy from one form to another. In the case of microsensors, the device typically converts a measured mechanical signal into an electrical signal. Microactuators move themselves or another component from one position or physical state to another.

Surprisingly, the micromachined version of a sensor usually outperforms a sensor made using the most precise macromachining techniques. There are numerous examples, such as pressure transducers and accelerometers. Not only is the performance of MEMS devices superior, but micromachining uses the same batch fabrication methods used in the integrated circuit industry. Thus, hundreds or even thousands of devices can be made on a single silicon wafer, and the cost per item becomes insignificant. Consequently, it is possible to achieve not only stellar device performance, but to do so at a comparatively low-cost level.

Microactuators are a more recent development for MEMS: microvalves for control of gas and liquid flows, optical switches and mirrors to redirect or modulate light beams, independently controlled micromirror arrays for digital projectors and displays, microresonators for capturing light and making it resonate at particular frequencies (such as in lasers or spectrometers), micropumps to develop positive fluid pressures in medical devices, microflaps to modulate airstreams on airfoils, and many more. And even though these microactuators are extremely small, they can cause effects at the macroscale level, performing mechanical feats far larger than their size would imply.

However, this is still very much a new frontier. When MEMS sensors, actuators, and structures can all be merged into a common silicon substrate along with integrated circuits (IC) for microelectronic systems, that is when the true potential of MEMS will be realized. Currently, electronics are fabricated using IC processes, and the MEMS components are fabricated using "micromachining" processes that selectively etch away parts of the silicon wafer or add new structural layers to form the mechanical and electromechanical devices. The present state-of-the-art usually involves a single discrete microsensor or microactuator integrated with the electronics. To merge with the electronics, the MEMS component is usually extracted from its silicon wafer and adhered, soldered, or fused to an IC, which may include a printed circuit board (PCB) or other components significantly larger in size than the MEMS device.

According to MEMSnet, "this vision of MEMS whereby microsensors, microactuators and microelectronics and other technologies, can be integrated onto a single microchip is expected to be **one of the most important technological breakthroughs of the future**. This will enable the development of smart products by augmenting the computational ability of microelectronics with the perception and control capabilities of microsensors and microactuators."

Soon, "smart implants" will be the norm, where ICs provide the "brains" of a system and MEMS components provide the "eyes" and "arms," allowing microsystems to sense, interpret, diagnose, and alter their environment. Already, MEMS lab-on-a-chip systems are revolutionizing many product categories by enabling complete diagnostics and assays to be performed in the field without the time or equipment required by traditional methods. Soon, this same technology will be implantable, providing immediate information to caregivers, or even spontaneous treatment to patients.

Nanotechnology is a popular buzzword these days. Dimensions are on the order of 10^{-9} m, whereas MEMS is micro at 10^{-6} m, with many dimensions on the millimeter scale. Nanotechnology is characterized by its ability to manipulate matter at the atomic or molecular level, and typically the behavior of the device *is different* than how it would behave at the micro- or macroscale. For instance, carbon nanotubes and needle arrays at the nanoscale can be arranged close enough together that their separation is smaller than the diameter of a water molecule; thus, a water molecule will roll across the surface and cannot penetrate—if applied to a fabric, it becomes waterproof and stainproof.

The important thing to remember is that nanotechnology is often not feasible without MEMS. In order to study nanoscale effects, a MEMS-scale "laboratory" is required. Further, the distinction between nanotechnology and MEMS is fluid. If even one active feature is smaller than 100 nm in size, it is by definition "nanotechnology."

There are two approaches for implementing nanodevices: top-down and bottom-up. In the top-down approach, devices and structures are made using many of the same techniques as used in MEMS except they are made smaller in size, normally by employing more advanced patterning and etching methods. The bottom-up approach usually involves deposition of layers, growing, or even self-assembly technologies.

Recently, many MEMS technologies are now dependent on nanotechnologies to create new products. For example, automobile SRS airbag accelerometers are manufactured using MEMS technology, but they can have long-term reliability issues due to stiction effects between the proof mass and the underlying substrate. A nanotechnology called self-assembled monolayer (SAM) coatings is now routinely used to treat the surfaces of the MEMS elements in order to prevent stiction effects from occurring.

Thus, MEMS and nanotechnology are part of an integral whole and the contributions of each should be considered when designing and fabricating your device.

How Do You Fabricate a MEMS Device?

As an example, let us say you want to create a MEMS strain gauge (microsensor). A strain gauge is a long "wire" that has been folded back and forth dozens of times for size convenience, with large solder tabs at each end for connecting to a circuit. When the wire is placed onto a specimen, as the specimen microscopically changes in length (under load and/or temperature), the "wire" changes in length, and this is measured as a change in resistance of the "wire." Typically, strain gauges are printed with conductive copper onto flexible substrates. Since silicon is a semiconductor, MEMS strain gauges are simply silicon "wires" that are etched in the folded shape. The microfabrication method may be different, but the strain gauge operates in exactly the same way.

For microfabrication, you need to think in terms of patterned layers. Your silicon wafer, usually 100–200 mm (4–8 inches) in diameter and 600 μm (0.024 inches) thick, can be purchased polished on one or both sides. Let us assume you want to use a single-sided wafer for this device, as you plan to make your strain gauge the full thickness of the wafer. The unpolished backside will provide some electrical isolation and make it easier to adhere your strain gauge to your specimen (Fig. 1.1).

The microfabrication process begins on your computer with a specialized CAD program (such as Layout Editor, https://www.layouteditor.org/) to create the layers you need to pattern your wafer. These layers represent each of the processing steps required to fully fabricate your device, such as the size and shape of the looped "wire" of your strain gauge, the soldering tabs, and the mote, with tethers around each strain gauge to allow you to easily break them free from the substrate.

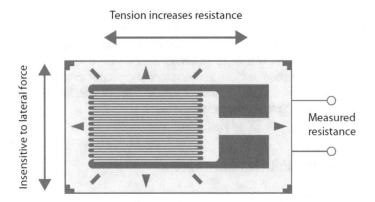

Fig. 1.1 Typical strain gauge

Fig. 1.2 Image of an
etched silicon wafer

Next, you cover your wafer with as many copies of the strain gauge as you can
fit. This can be anywhere from a hundred to a thousand, depending on how densely
you can package them and the diameter of the wafer you are using. Then, you either
make or send out your patterns to make masks for each layer. The mask, made of
glass, can either be a positive mask or a negative mask, depending on whether or not
you want to expose the feature to be etched or expose the negative space *around* the
feature to be etched (Fig. 1.2).

Once you are ready with your wafers purchased and your mask patterns designed
and fabricated, you need to gown up and enter the cleanroom. Unlike most macroscale fabrication, microscale work is done in a clean, particulate-controlled environment that must be protected from dust that you could bring in on your hair,
clothing, shoes, or even electronic devices. So, covering all of your entire self and
cleaning your electronics is required.

There are over a dozen national MEMS labs that are available for public use. How to find these facilities and get trained to use them is covered in a later chapter. Let us assume for now that you have identified a lab and have the proper training. Your first stop in the lab will be to passivate your wafer to remove any free electrons. Then, you will need to clean and spin-dry your wafer to remove any dust or ash that may have accumulated on the surface. It is impossible to tell if your wafer is contaminated, so you will always assume that it is.

Overview of the Fabrication Process

Wafers are usually patterned with high-intensity light passing through the masks you have prepared. This is called photolithography. For this process, you first spin coat on a micro-thin layer of a photosensitive film, bake it to harden, and then place your mask for the first layer into a photolithography machine between the light source and your coated wafer. A few seconds of light exposure and the parts of the film that were exposed by the mask are degraded in a way that allows the material to be washed away in a chemical bath, revealing the silicon wafer again. As with the microscopic contaminant, you quite often cannot see the pattern that you have created on the wafer with your naked eye. You would have to look at it under a microscope, but this is not advised, as you can potentially expose your pattern to harmful light, even if you use a filter. So, you again assume that all is well and proceed to etching or build-up step. Let us use an etching and material removal process for our strain gauge example.

Etching can be either wet or dry. Wet etching uses a chemical bath to eat away at the silicon wafer wherever it is exposed. The shape and depth of the etch depend on several factors, including obvious ones like time and the chemical used, and not so obvious ones like the silicon crystal orientation in your chosen wafer. If the wafer has Miller index orientation of <100>, that means the face of the crystal lattice is normal to the face of the wafer. The other axis orientation is determined by a flat on the circumference of the wafer and is used for alignment purposes. During etching, the chemicals eat away along some crystal planes more rapidly than others, so if your goal is to produce a trapezoidal feature, you will choose one orientation, perhaps a < 111>. If you want more rectilinear walls, you will choose a different one. Not to overly complicate the choices available to you, but you can also choose what type of doping you want, which determines whether there is an excess or lack of electrons available. None of these choices, however, are something you need to determine by yourself. All facilities have staff that can help you choose the most appropriate wafer for your project.

Dry etching involves bombarding the surface with ions. The area protected by the film is not etched. This is called reactive ion etching (RIE) or deep reactive ion etching (DRIE) for greater depths. The advantages of dry etching are numerous. First, it is much easier to get straight walls on your features. Second, you can etch much deeper (on the order of 40 μm) and maintain very tight tolerances of 1–2 μm.

Third, there are less variables to consider, even though the setup is more complex, so it is easier to control the outcome and maintain consistency between setups.

Other steps may involve adding material, such as metal, for your strain gauge's soldering tabs. This is called vapor deposition. Although the material is deposited uniformly across the wafer, that which is on top of the film is washed away, leaving metal adhered only where desired.

You repeat the steps of spin coat on a film-then pattern-then etch or deposit-then rinse for as many mask layers as you have in your design. This can be as few as three or four or as many as 40 or more. Along the way, you may choose to microscopically evaluate or measure a feature to ensure it meets your requirements. Some processes can take several hours, others just minutes, and it is not uncommon for a single wafer to take multiple days to complete. Eventually, you are done. You take your wafer out of the cleanroom and either break or dice your devices out of the wafer. They may be simple mechanical devices like our strain gauge sensor, or they may be complex microactuators integrated with some of their own circuitry, paired with surface-mount soldered components and assembled onto printed circuit boards. The possibilities are limitless.

Learning About What Is Possible

Preparing to undertake a first project in MEMS can be daunting. As a new user, you have no overall sense of the technology or what is possible, so it is important to familiarize yourself with devices that are currently marketed and to research concepts suitable to your application. An Internet search on "MEMS" will reveal endless results, so be more focused and specific. If you are wanting to sense a certain substance, say glucose, then that should be your research target. It turns out glucose can be sensed in blood, within certain cells, in saliva, in tears, in urine, in breath, and more.

If you decide your target sensing method is glucose in one's breath, that will lead to much different MEMS sensor designs than if your target method is in saliva. For breath, you may want a chemical or biological film that is "activated" in the presence of glucose. The activation can lead to a color change, phosphorescence, physical weighting of a thin beam causing electrical contact, a change in resistivity, or something else.

You will likely not know all of the possible solutions, which is where the expertise of the staff at the lab is invaluable. It is perfectly acceptable to bring in your project with as little as "glucose sensing in one's breath" and discuss how you would want it to be used clinically or by the end user. The staff will hold a design review with you and brainstorm on possibilities, explaining how each could be implemented using the equipment in the lab.

Once you have decided upon a preferred implementation, they will work with you to learn all the skills needed to fabricate the sensor yourself, or connect you with someone who is willing to collaborate and take on your project for a fee.

Skills Required

No prior knowledge with microfabrication is required to be successful creating a MEMS device. After reading this book, you will be able to do it, regardless of your background. In general, technical people, especially machinists and engineers, readily pick up on the methodology and quickly understand the operation of the equipment. Machinists will recognize the similarity between macroscale machining and microfabrication, as material removal is material removal. New tools are required, nothing more. Engineers, irrespective of discipline, will be able to relate to the underlying physical and electrical principles, and since they are accustomed to using technology, this will feel like an extension of that base knowledge.

However, a technical background is not necessary. Anyone with an interest can learn how to design and fabricate MEMS devices. Step-by-step training will be provided by the lab, and there is literally no question you could ask that they have not answered before. So, take your entrepreneurial spirit and feel free to dive right in!

Starting at Larger Scale

When you have a design concept for your MEMS device, remember that it is always useful to do a mock-up or prototype at a larger scale. Many things that can be micro-fabricated in silicon with etching techniques can also be laser cut from thin sheet metal, or machined from wood or plastic. Unlike nanotechnology, microscale devices behave approximately the same at macroscale as they do at microscale. Also, MEMS can be done just as easily at the millimeter scale, so spend some time upfront constructing a prototype. Attach it to a full-size circuit board with off-the-shelf components and do some testing. You will learn a lot about your design, preventing a costly investment in a microscopic device that is difficult to test and in the end does not function correctly.

Slowly size down to your ultimate goal. After functional testing, step down to a MEMS-scale device, but with a miniaturized circuit board that uses surface-mount off-the-shelf components. The sensor system shown on the back cover of this book is an interim design of my device. The sensor is to scale, but the circuit is "larger" and uses standard resistors, capacitors, and op-amps to create a signal amplifier. Before this stage, I microfabricated only the sensor at a slightly larger size, and as shown in Fig. 1.3, it was hardwired to a crude circuit board (with insulated cables to reduce noise). The cables were attached to benchtop equipment, such as a function generator, power supply, and oscilloscope, so I could characterize the sensor's output.

One of the main challenges you will face when working with MEMS devices is scale. Often, you cannot even see the features of your device without a microscope. Everything is more difficult—holding the device, releasing it from the wafer, adhering it to your specimen or packaging, soldering to it for electrical connectivity,

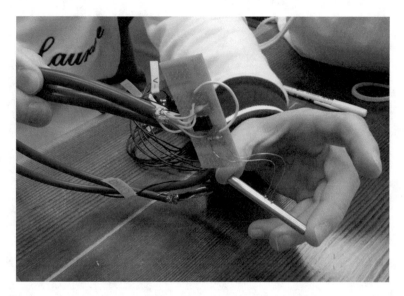

Fig. 1.3 Early sensor prototype with external circuit board

amplifying its tiny output to a level you can read, and so on. The key to success is thus to face one challenge at a time. Make sure your MEMS device works at a macroscale and then microfabricate only it and test that. When you are satisfied with those results, then start reducing other parts of the system. That way, when something goes wrong, you will know where to start looking (likely while trying to hold things and use the microscope).

Chapter 2
NNCI

The National Nanotechnology Coordinated Infrastructure (NNCI) is a network of 16 institutions across the United States. Most are affiliated with one or more universities. For instance, the one I have used is at the University of Washington, but Oregon State University is a partner institute and has additional equipment and resources available. The following map shows where the sites are located. Equipment available at these facilities varies, so review the website for more detailed information at www.nnci.net, where you can find a list of over 2000 pieces of equipment, a list of experts at each facility, and more.

Several of the labs are conveniently located near existing orthopedic and medical device manufacturing hubs, such as Austin, Boston, Minneapolis, Philadelphia, Portland, and San Diego, making them accessible to your current workforce (Fig. 2.1).

Structure of Organization

The NNCI Coordinating Office is at Georgia Tech University (https://www.nnci.net/nnci-coordinating-office) and is led by Prof. Oliver Brand who serves as director. Dr. David Gottfried serves as deputy director and oversees the day-to-day operations, assisted by Program Manager Ms. Amy Duke. In addition, there are three associate directors who manage the network activities in specific areas: Dr. Nancy Healy coordinates the NNCI education and outreach programs, Prof. Jameson Wetmore coordinates the societal and ethical implications (SEI) activities, and Prof. Azad Naeemi coordinates the computational activities and facilitates interactions with nanoHUB/NCN at Purdue University. As leadership roles may change over time, please refer to the website for the most current information.

These staff members are guided by an executive committee which includes the 16 NNCI site directors. The executive committee meets monthly via teleconference and annually in person at the NNCI Conference. In addition, there is an external

© Springer Nature Switzerland AG 2019
D. Munro, *DIY MEMS*, https://doi.org/10.1007/978-3-030-33073-6_2

Fig. 2.1 NNCI facilities map

advisory board (EAB) with members representing industry, academia, government, education and outreach, the SEI ethics subcommittee, nanoscience, and nanoengineering. The EAB meets in person as part of the NNCI Conference, with additional communications as determined by the EAB.

In addition, there are several subcommittees that discuss high-level issues related to the NNCI network as a whole, along with several working groups, composed of staff members from the NNCI sites, who share and develop best practices for site and network operations, technical areas, research areas, and education and outreach.

Other roles of the coordinating office include communicating with the National Science Foundation (NSF), maintenance of the NNCI website, collection and analysis of monthly site usage metrics, organization of the NNCI Annual Conference, and preparation of the NNCI Annual Report.

When/How Established

One of the major objectives of the National Nanotechnology Initiative (NNI), established in 2000 to coalesce the nanoscale research and development activities of more than 20 federal agencies, was to create user facilities and networks that were available to all. Recognizing that nanoscale science and engineering often requires expensive equipment and specialized expertise, only the largest research-intensive universities and national laboratories could afford to operate them. To alleviate this discrepancy for both smaller academic institutions and commercial enterprises, the NSF developed a new model of support now known as NNCI.

The NSF has been trying to promote a network that works for individual users and small companies, but for over 40 years, only academics at the sponsored university really knew this resource was available to them. Initially, there was the National Nanotechnology Facility at Cornell University (1977–1993), followed by the National Nanotechnology Users Network (NNUN, 1993–2003), and then the National Nanotechnology Infrastructure Network (NNIN, 2004–2015), which provided specialized nanotechnology resources to all researchers who needed them. The National Nanotechnology Coordinated Infrastructure (NNCI), established in 2015, is the latest version of this national resource.

The NNCI site awards are the result of a competition conducted by the NSF, which was generated as a result of input from the science and engineering community. Over 50 proposals from potential NNCI sites were submitted, resulting in 16 awards. The total level of funding for NNCI sites and the NNIN Coordinating Office is approximately $16 million annually. The initial cooperative agreements are for a period of 5 years with possible renewal for 5 years. I participated in the proposal put together by the Washington Nanofabrication Facility (WNF), so I feel a small measure of pride that by being an outside user, I helped them secure an award.

Research done at the various NNCI facilities is incredibly broad, with applications in electronics, materials, biomedicine, energy, geosciences, environmental sciences, consumer products, and many more. Each site is designed to accommodate research in a wide variety of materials and processes for both devices and systems for both top-down and bottom-up approaches to micro−/nanoscale science and engineering.

All of the NNCI facilities (many sites have partners and multiple locations) can be accessed by students and professionals from around the country and globally. The facilities within NNCI are meant to support both research and development— for academic research, as well as product development, technology and innovation, and commercialization for start-ups and larger, more established companies.

That is where this book comes in. Although the mission of NNCI is to allow for industry use, that is not the reality. Most facilities still have the vast majority of their users come from the host university. They get a small percentage of users from other academic institutions, and they support some community outreach in the schools, but it is virtually unknown to industry that they are welcome to use these facilities, too. So let us change that!

How to Become an NNCI User

The following information is also available in more detail at https://www.nnci.net/becoming-user.

Each NNCI site is an independent operating unit and has developed its own procedures for new user access, which you can find those on each of the individual site pages. The complete contact information is provided in a later chapter for all 16 sites. Although there are many different pathways for how to collaborate success-fully on a project, guidance for the overall process is provided at the website noted above and summarized as follows:

1. If you do not already have a specific site in mind or are not sure which site might be most appropriate, you can fill out the contact form for the "New User Gateway" and an expert will reply to assist you. After reading this book, however, you will have a good idea of what each site offers and will be armed with better informa-tion before using the gateway.
2. Contact one or more NNCI sites directly to discuss your project. A "New User Contact" is listed on each site page, and someone will reply to help you assess the technical feasibility of the project and the scope of your work. You may be asked to prepare a brief written statement of work, and discussions with techni-cal staff at the site might be needed as well.
3. After initial discussions, the site contact will help you establish a user agreement with the site, get you the credentials needed to begin orientation and tool train-ing, and coordinate necessary support toward the completion of your project.
4. The fee structure, invoicing, and payment process is unique to each site and should be discussed before any agreement is established. For most sites, this information is available online, and a sample fee structure is provided for WNF in a later chapter.
5. Once any paperwork is completed, your work at the site (or completion of the remote project) will be scheduled at a mutually agreeable time. In most cases, work can start within a matter of weeks. The site will arrange for appropriate training and project supervision by experienced staff.
6. If you need it, feel free to contact the NNCI Coordinating Office for additional assistance.

NNCI New User Gateway

The online form for the New User Gateway can be found at https://www.nnci.net/contact/get_started. The fields to be completed are as follows:

GETTING STARTED ON BECOMING AN NNCI USER
Are you a new user trying to get started? Fill out the handy form below, and one of our representatives will contact you.
>   FIRST NAME
>   LAST NAME
>   EMAIL ADDRESS
>   WHERE ARE YOU

- What school or company are you affiliated with?
- Optional choice of your preferred site.

- Which group best describes you?
- Please choose the category that you are interested in researching?

PROVIDE YOUR QUESTIONS OR COMMENTS HERE.

NNCO, Webinars, and Resources

The National Nanotechnology Coordination Office (NNCO) is a support service group of the US National Nanotechnology Initiative (https://nano.gov). It offers quarterly newsletters, an ongoing webinar series, and information about upcoming events related to NNCI and nanotechnology. Tabs on the webpage link to:

- Nano 101.
- About the NNI.
- Networks and Communities.
- Publications.
- Commercialization.
- R&D Infrastructure.
- Educational Resources.
- Communications.
- Events.

The Commercialization tab discusses FAQs for business, funding opportunities, nanotech challenges, diversity programs, and entrepreneurship networks. This is a good place for industry users to gather more information about opportunities.

In addition, the website has links to several national R&D user facilities under the link https://www.nano.gov/userfacilities:

- Center for Nanoscale Materials, Argonne National Laboratory.
- Center for Functional Nanomaterials, Brookhaven National Laboratory.
- Molecular Foundry, Lawrence Berkeley National Laboratory.
- Center for Nanophase Materials Sciences, Oak Ridge National Laboratory.
- Center for Integrated Nanotechnologies at Sandia National Laboratories and Los Alamos National Laboratory.
- The National Cancer Institute's Nanotechnology Characterization Laboratory (NCL).
- The National Institute of Standards and Technology's Center for Nanoscale Science and Technology (CNST).
- The National Science Foundation's Nanotechnology Coordinated Infrastructure (NNCI), which is the focus of this book.

Sample New User Agreement with a Site

The following is excerpted from WNF's Industry User Agreement, for persons who wanted to do the work themselves in the facility.

This Facility Use Agreement (Agreement) is between the University of Washington (UW) and the user identified below ("LAB USER"), who is either a non-UW student or an employee of or an independent consultant under contract to the institution identified below (the "INSTITUTION"), regarding the LAB USER's shared use of the Washington Nanofabrication Facility (WNF) operating in Fluke Hall and/or Molecular Analysis Facility (MAF) operating in the Molecular Engineering and Science Building located on the principal campus of the University of Washington in Seattle, Washington. The WNF together with the MAF form an integrated research laboratory funded, in part, by the National Science Foundation (NSF) though the National Nanotechnology Coordinated Infrastructure (NNCI).

Laboratory Policy: The LAB USER agrees to abide by all laboratory policies as outlined in:

• The WNF user manual, posted at: https://www.wnf.uw.edu/docs/WNF-User-Manual.pdf

• The MAF policies, posted at: http://www.moles.washington.edu/maf/access/userpolicies/

Although the WNF and MAF provide general safety courses and training on the safe use of specific equipment, the LAB USER assumes responsibility to plan and perform work in such a way as to ensure his or her own personal safety as well as the safety of others in the facility.

Fees: The INSTITUTION acknowledges responsibility for purchases, materials costs, and lab fees incurred by the LAB USER in his or her use of the WNF and/or the MAF. A listing of the current fees can be found at:

• WNF tools: https://www.wnf.uw.edu/docs/WNF-Rates.pdf

• MAF tools: http://www.moles.washington.edu/maf/

Any future fee changes in excess of 10% will be made with at least 90 days' notice. The INSTITUTION is responsible for promptly notifying the joint WNF/MAF facility operations manager in writing if a LAB USER ceases to be a student, or employee of, or under contract to the INSTITUTION and is responsible for all fees and costs by such LAB USER until such time as notice is received by the facility operations manager.

Nonpayment of fees within sixty (60) days of receipt of invoice by the INSTITUTION will subject the INSTITUTION to termination of laboratory access with ten (10) days' notice to the INSTITUTION. Finance charges at a periodic rate of 1% per month or 12% per year shall be added to balances past due over thirty (30) days. The INSTITUTION acknowledges it will be financially liable for equipment or other property damage if it is found to result from negligence or violation by the LAB USER of standard WNF/MAF policies and procedures. Checks, payable to the University of Washington, should be mailed to: University of Washington, Invoice Receivables, PO Box 94,224, Seattle, WA 98124.

Limits on Use: The WNF/MAF is a community of professional and student researchers; courteous, professional, responsible behavior is required at all times. Access to WNF/MAF will not be permitted until such time as the LAB USER has returned an executed User Agreement Form, the User Billing & Information Form, is covered by a Reviewed Research Project that is on file with and approved by the WNF and/or MAF lab managers, and has completed the mandatory orientation and safety classes for each laboratory tool and process the LAB USER will be using. New training is required for users who are inactive for more than one quarter or when deemed necessary by WNF/MAF staff. Access to WNF/MAF is a privilege and may be suspended, restricted, or have conditions placed upon it by UW at any time and for any reason at the discretion of the laboratory manager. Use of WNF/MAF is limited to research and development as described in the Reviewed Research Project; work outside its scope requires submission of an additional project proposal for review and approval by the WNF/MAF lab managers. Processes with the potential to significantly affect the research of other users or the general operation of the lab are not allowed (e.g., no proprietary chemicals may be brought into the lab). A LAB USER'S access to WNF and MAF is contingent upon his or her continuing affiliation with the INSTITUTION. If the LAB USER's affiliation with the INSTITUTION ends, his or her access to WNF/MAF will terminate until such time as the LAB USER has a new or renewed affiliation with an INSTITUTION and a new User Agreement Form and User Billing & Information Form is submitted. Sharing of access cards with other users or allowing unauthorized access to the facility is strictly prohibited and may be grounds for terminating facility access. Non-authorized persons are prohibited from accompanying, observing, or helping users at work unless specifically approved by the laboratory staff.

Research, Intellectual, and Personal Property Rights: The LAB USER and INSTITUTION acknowledge responsibility for their own research and that WNF/MAF do not in any way warrant or assure project success.

The LAB USER and INSTITUTION further acknowledge responsibility for their personal and intellectual property. WNF provides limited, unsecured storage as a courtesy and makes no guarantees against unauthorized access by non-INSTITUTION individuals. If a LAB USER ceases to be a student or employee of, or under contract to, the INSTITUTION, or if the relationship between the INSTITUTION and WNF/MAF is terminated, the INSTITUTION is then responsible for removing any personal property within sixty (60) days, or it may be disposed of at the discretion of WNF/MAF staff. In addition to the LAB USER's use of the lab, the INSTITUTION may arrange for the participation of UW personnel for the conduct of proprietary research. All such arrangements shall be made under separate written agreement with the UW.

NNCI Program Requirements: Submission of an annual report for active project is required by NSF as a provision of the NNCI program. The LAB USER and INSTITUTION agree to provide a project title and brief description of work accomplished during the year; the report should not contain sensitive information, as it may be used in presentations to illustrate the range of research topics at WNF/MAF. The INSTITUTION also acknowledges that its identity may be made public in presentations and other materials describing WNF/MAF and the NNCI. The INSTITUTION further agrees that, where appropriate, the WNF/MAF and the NNCI will be acknowledged in any of its sponsored publications or presentations, resulting from substantive work performed at these facilities. A suggested acknowledgment is: "Part of this work was conducted at the Washington Nanofabrication Facility / Molecular Analysis Facility, a National Nanotechnology Coordinated Infrastructure (NNCI) site at the University of Washington, which is supported in part by funds from the National Science Foundation (awards 1542101, 1337840 and 0335765), the National Institutes of Health, the Molecular Engineering & Sciences Institute, the Clean Energy Institute, the Washington Research Foundation, the M. J. Murdock Charitable Trust, Altatech, ClassOne Technology, GCE Market, Google and SPTS." Also, electronic copies of all publications that contain work done at WNF/MAF must be sent to the WNF/MAF associate director.

Liability: The INSTITUTION acknowledges responsibility and liability for the acts and negligence of its employees and agents and maintains health, accident, and workers' compensation insurance for the LAB USER while he or she is working at WNF/MAF. The LAB USER and the INSTITUTION understand that use of WNF/MAF may involve exposure to potentially hazardous conditions including, but not limited to, chemical, mechanical, electrical, thermal, and radiation hazards. INSTITUTION's health and accident insurance coverage shall cover problems related to these hazards. The parties agree that the relationship between the parties established by this agreement does not constitute a partnership, joint venture, agency, or contract of employment of any kind between them and that nothing herein shall be interpreted as establishing any form of exclusive relationship between the parties. The LAB USER and the INSTITUTION shall release, hold harmless, and indemnify the University of Washington, its regents, officers, agents, employees, and students from any and all claims, damages, costs (including reasonable attorney fees), and liabilities arising out of the LAB USER's use of the WNF/MAF facilities other than such as results from the gross or sole negligence of the University of Washington, its regents, employees, officers, agents, students, or representatives under this agreement. Neither party shall have any liability of any kind to the other party for any indirect damages, including, but not limited to, lost profits, lost revenues, or loss of use.

Term and Termination: Subject to its other provisions, this agreement shall commence on the start date below and shall automatically renew annually on July 1 of each year unless previously terminated. Either UW or INSTITUTION may terminate this agreement by giving thirty (30) days' prior written notice to the other. UW may terminate the agreement by giving ten (10) days' notice in the event of (i) failure to timely pay charges as noted above, or (ii) violation of rules or operating procedures established in the "Washington Nanofabrication Laboratory User Manual" or "Molecular Analysis Facility Policies." In the event of such termination, the INSTITUTION will only be liable for facility use costs incurred up to the date of termination. No use of the WNF/MAF, Fluke Hall or MolES building, or equipment shall extend beyond the termination of this agreement without prior written approval of UW.

Dispute Resolution: The parties hereby consent to and accept the exclusive juris-diction and venue of the Superior Court of King County, Seattle Division, Washington, in any dispute arising under this agreement. The rights and obligations of the parties under this agreement shall be governed by the laws of the State of Washington. In the event an action is commenced to enforce a party's rights under this agreement, the prevailing party in such action shall be entitled to recover its reasonable costs and attorney's fees, as determined by a court in conjunction with such legal proceedings. If any of the provisions of this agreement shall be deter-mined to be invalid, illegal, or unenforceable by a court, such provision shall be automatically reformed and construed so as to be valid, legal, and enforceable to the maximum extent permitted by applicable law while preserving its original intent, and the other provisions shall remain in full force and effect.

This user agreement is provided only as a sample and you should expect there to be differences from site to site. In general, the guidelines for facility use are fairly standard and not unlike agreements with other organizations.

Chapter 3
Working in a Cleanroom

A cleanroom is designed to control particulate debris—things like dust, lint, dandruff, hair, ash, and powder. The cleanroom is not necessarily sterile and often contains numerous chemical contaminants and hazards. In fact, much of the protection worn in a microfabrication cleanroom is for your safety against the hazards present in the environment—gloves, face shields, full skin coverage suits, and more provide you protection from exposure. One of the hardest habits for me to overcome was my desire to shake hands with others while in the cleanroom. This is simply not done, as it can transfer toxic chemicals from one person to another. In addition, there are no doorknobs for the same reason.

Cleanrooms are designated by classes—the smaller the number, the less particulates are found in the environs. There are US Federal and ISO international classifications, and all refer to the maximum number of particles larger than a certain size per a certain volume. For MEMS work, most cleanrooms are certified Class 10,000 and follow protocols that bring them closer to Class 1000. For Class 10,000, the maximum number of particles less than or equal to 0.5 μm is 10,000 per cubic foot, and it also allows for 70 particles less than or equal to 5 μm in that same volume. For reference, a human hair averages 75 μm in diameter. In ISO terminology, a Class 10,000 cleanroom correlates to an ISO7 cleanroom, where the number of particulates is measured for a volume in cubic meters (Fig. 3.1).

Protocol for Cleanroom Use

Each facility will have governing protocols for all aspects of lab use, which covers important aspects of safety, chemical use, cleanroom gowning, and more. At the WNF, this document is called the user manual and can be found at https://www.wnf.washington.edu/docs/WNF-User-Manual.pdf. Much of the work detailed in the WNF user manuals was researched and compiled over many months by Dr. Andrew Lingley, now the Manager and Research Engineer at the Montana Microfabrication

© Springer Nature Switzerland AG 2019
D. Munro, *DIY MEMS*, https://doi.org/10.1007/978-3-030-33073-6_3

Fig. 3.1 Example of full
cleanroom gowning

Facility at Montana State University. I want to thank him for providing this valuable information to all of us. There are many additional aspects of lab use covered in this comprehensive manual, but the key ones to understand as a new user of a cleanroom are facility safety, chemical use, and gowning. All of this will be covered fully in training (and you will be tested on it) before you are allowed to enter the facility.

Facility Safety

Before entering a cleanroom facility, you need to be aware of the potential hazards and what to do in case of different types of emergencies. First of all, always sign in and out of the facility so that other uses and emergency personnel are aware you are inside. Most facilities have many rooms and countless blind spots, so logging in is critical. In case of emergency, you should know where the exits and fire alarms are located, and you should know where to meet if you need to evacuate the building. In some instances, it will be dangerous to cross a spill or breathe the fumes involved, so be sure you know multiple routes to exit the building. You should also know where phones are located and what the emergency service phone numbers are.

Chemical Use

By its very nature, MEMS fabrication requires the use of a lot of chemicals. Wafers are cleaned, coated, wet etched, rinsed, and more using various chemicals. Some commonly used chemicals are extremely dangerous, such as hydrofluoric acid (HF), which is used for some types of etching. It differs from other acids because the fluoride ion readily penetrates the skin, causing destruction of deep tissue layers, including bone. It is colorless and creates fumes, so it must be used inside a mostly closed fume hood with additional safety equipment, including a face shield worn over your safety glasses, mask, and head covering, a chemical apron, and a third set of heavy gloves.

Other chemicals are flammable, such as acetone, which is a colorless liquid used for cleaning. Acids, bases, flammable liquids, and other chemicals are stored separately in well-marked cabinets. You may have to dispense a small amount into a beaker or squirt bottle before use, and if so, that secondary container must be labeled with its contents, the date, your name, and when it will be properly disposed.

Nothing is more important than your safety, so during your training or even after you have been using the lab for some time, ask someone if you have any doubts or questions before using or disposing of a chemical to make sure you are doing so correctly.

Cleanroom Gowning

Entering a cleanroom is an involved procedure. My advice is to make sure you have taken care of your personal needs before doing so, as taking a break to use the restroom, get a snack, or make a phone call is tedious. Every site will have its own specific procedures, but they are all similar in a broad sense. The following are the procedures for WNF, which can be read in detail in the user manual.

Visitors

Visitors are not allowed in lab spaces without approval. If you want to bring visitors into lab spaces, you must submit an online Visitor Request Form (https://www.wnf. washington.edu/lab-user-portal/visitor-request/) prior to the visit. For each approved visitor you will be assessed a fee to cover cleanroom consumables. An active WNF user must escort each visitor during his or her entire time in laboratory spaces and is responsible for the visitor's actions and safety. Visitors are to abide by all safety measures expected of laboratory users and are not allowed to operate any laboratory equipment or perform any chemical operations. Visitors are not allowed in the facility after normal hours or as buddies to meet the buddy system requirement.

Dress Code

The WNF dress code applies to all laboratory areas, cleanroom and non-cleanroom. It is based on industry-wide best practices for working in areas containing hazardous materials and emphasizes.

covering as much skin as possible to contain sloughed skin particles. You will be denied access to laboratory spaces if you are not dressed according to the following rules.

Hygiene

You and your clothing should be clean (i.e., free of dust or dirt) before entering lab spaces. Avoid clothing that sheds fibers such as wool, fur, fake fur, and mohair. Dirty or shedding clothes contaminate everyone's work, even in non-cleanroom spaces. Makeup and cosmetics are not allowed in lab spaces.

Shoes

You must wear shoes that completely enclose the heel, toes, and top of your feet. Socks or stockings are required. Sandals, open-weave shoes, or shoes that expose the top of the foot are not allowed. High heels and deep-treaded shoes that hold mud or dirt are also not allowed. Despite the substantial wet weather in the Pacific Northwest, your shoes should be clean and dry before entering the labs. It is preferable that you bring a clean, dry change of dedicated lab shoes that you do not wear outside. It is imperative that you avoid tracking mud, dirt, and leaves into the labs.

Pants

You must wear long pants that run from your shirt to your ankles. Shorts, short pants, capris, skirts, and dresses are not allowed. During warmer weather, you may bring with you a pair of lightweight hospital-scrub-style pants to wear over your shorts. You can purchase inexpensive scrub pants from the WNF staff for this purpose. You can either purchase a pair of scrub pants from WNF staff or need to return home to change into acceptable attire.

Shirts

Your shirt must cover your shoulders and reach from the top of your arms to your pants. Tank tops, halter-tops, and spaghetti strap tops are not acceptable.

Safety Glasses

Safety glasses must be worn at all times in all labs. The WNF provides safety glasses, or you may choose to purchase your own glasses, provided that they meet ANSI Z87.1–2003 standards. You are allowed to remove your safety glasses when using optical microscopes, but remember to put them back on when you step away from the microscope. Safety glasses are not acceptable for chemical protection; face shields must be worn during chemical use. Please refer to the personal protective equipment (PPE) section for details in chemical protection protocols.

Contact Lenses

Consistent with recent recommendations from the American Chemical Society, contact lenses are allowed in WNF laboratories, provided that safety glasses are also worn at all times. In the case of an eye exposure emergency, rinse at the emergency eyewash station with contacts in place, and remove them while flushing.

Gowning Procedures

Proper gowning is important to maintain garment and facility cleanliness. Gowning will be demonstrated for you during your cleanroom orientation.

To enter the cleanroom, swipe your keycard to unlock the gowning room door. Even if someone else opens the door for you, you must also swipe your keycard to indicate your entry into the cleanroom. CORAL tracks the real-time user list for emergency response accountability. It is imperative that you swipe out of the lab when you leave; otherwise, EMS personnel will assume you are in the lab in an emergency. If your keycard does not unlock the door, you may not enter the cleanroom. If you have completed all access requirements and your access card does not work, please see or email lab administration in Fluke Hall Room 215 to troubleshoot access issues. Entering the cleanroom without swiping your keycard or when your keycard is disabled is a violation of both lab and university policy.

- Before entering the cleanroom, make certain you meet the dress code requirements.
- Before stepping beyond the first bench, put on a pair of blue shoe covers; a bouffant hair net, enclosing as much of your hair as possible; and a first pair of nitrile cleanroom gloves.
- If you do not have a cleanroom suit already on a hanger, select a hood, cloth face veil, coverall, and pair of boots in your appropriate size from the shelves.
- Put on the cleanroom hood with the seams facing inward, and then snap the cloth face veil inside the hood. Some of the hoods are packaged inside out, so check that you are putting the hood on correctly.
- Verify placement of the face veil using the mirror on the wall. Some of the hoods have a pair of snaps under the chin. There are multiple snaps on the hoods and face veils so you can adjust and find positions that provide maximum cleanliness and comfort. The face veil needs to completely cover your mouth and nose, resting on the bridge of your nose where your safety glasses rest.
- Place your safety glasses outside of your hood and on the bridge of your nose where the face veil rests.
- Put on the cleanroom coverall suit; do not to drag the suit on the floor in the process. Hold the sleeves in your hand while putting your feet into the suit to prevent the sleeves from touching the floor. Your first pair of gloves should be tucked under the cuffs of the suit.

- Tuck the bottom of the hood into the suit, zip up your suit, and snap the top snap on your suit. Use the mirror on the wall to verify that the hood is correctly positioned inside the cleanroom suit and that both your mouth and nose are covered.
- Sit on the second bench and put on your white cleanroom boots, tucking the legs of the suit into the boot. Connect the strap across the top of the foot and tighten it snugly. Snap the top of the boot to the back of the suit leg. Avoid putting your clean white cleanroom boot down on the gowning room side of the bench, and avoid putting your blue bootie-covered foot down on the cleanroom side of the bench.
- Put another pair of gloves on over the top of the gloves you initially put on (double-glove). These should extend outside the sleeves of the coverall.
- Using an IPA (isopropyl alcohol) squirt bottle, moisten a cleanroom wipe and wipe down all the items you bring into the cleanroom (such as laptop or cell phone).
- Once inside the cleanroom, do not open or unzip your cleanroom suit. If you need to access something within your suit (e.g., cell phone), go into a gray area to do so.

Personal Effects Storage

While you are working in the labs or cleanroom, store all personal items, such as coats, knapsacks, bicycle helmets, and books, in the alcove just opposite the cleanroom entry door, near the central restrooms. Do not bring any of these items into the gowning room or cleanroom. This is a short-term storage; items left for prolonged periods may be disposed or reclaimed.

Cleanroom Item/Activity Restrictions

People and the items they bring into the cleanroom are primary sources of particulate contamination. In addition to the table above that applies to all WNF lab spaces, Table 3.1 lists items that can and cannot be brought into the cleanroom. Minimize contamination by only bringing items necessary for your research into the cleanroom.

Cleanroom Protocol

All users must help maintain the integrity, usability, and effectiveness of the cleanroom. Even if your project is not sensitive to particulates or other contamination, you must follow all cleanroom protocols at all times.

Table 3.1 Partial list of prohibited and allowed items/activities

Partial list of prohibited items/activities	Allowed items/activities
Cardboard, fiberboard, wooden containers	Plastic items, plastic boxes, plastic containers
Paper, paper notebooks, books, magazines, etc.	Cleanroom paper, cleanroom notebooks
Pencils, erasers	Pens
Over-the-ear headphones	Earbud headphones (keep at low volume)
Hats, coats, scarves, bags, backpacks, etc.	Over-the-ear headphones if fully under hood (at low volume)
Makeup, cosmetics	Laptops, e-readers
Running or jogging	Cell phones
Food or drink, gum, cough drops, mints, etc.	Cameras (no flash in yellow rooms)
Smoking	Wafer handling tools, supplies
Offensive or obscene materials or media	

- If your gloves are torn, soiled, or otherwise contaminated, immediately remove the outer pair and put on new gloves.
- Never touch doorknobs, telephones, equipment controls, microscopes, or other common objects with contaminated gloves. Cross contamination can permanently damage equipment and expose other users to chemical hazards.
- Tacky mats are placed throughout the lab to reduce airborne particulates. Do not step over or bypass the tacky mats.
- Do not leave items strewn about the lab. Use your assigned dry box storage space. Items left in the lab will be moved to the lab lost-and-found, in the east gray space. After 1 week, unclaimed items will be discarded or reclaimed.
- If you need something from under your suit, do not unzip your suit in the cleanroom. Move to a gray area (gowning room or maintenance/storage chase) to open your suit. If you intend to use the retrieved item in the cleanroom space, it must be wiped with IPA.
- Do not sit or lean on equipment or tables.
- Do not shake hands in the cleanroom.
- Use all materials (e.g., wipes and chemicals) sparingly to keep costs down.

Wafer Handling

Cassettes

Whenever possible use cassettes to carry and process your wafers, except in cases where using a cassette would result in excessive and unnecessary chemical usage. Although you can orient wafers in your cassette however you please, they are generally positioned starting from the H-bar (horizontal bar) with the polished surface facing away from the H-bar. Cassette-to-cassette transfers are the easiest method to transfer a large number of wafers. First, set the filled cassette on a flat surface, then flip an empty receiving cassette upside down, and mate the two cassettes using the dimples and holes. Grab both cassettes, compress them together, and then slowly tilt both until the wafers roll from the donor cassette to the receiving cassette.

Tweezers

Only handle wafers with wafer tweezers. Only touch tweezer handles; do not touch the shovel or pincer (gripping end). Ensure your tweezers are compatible with the chemicals your processing requires. Use care to avoid gouging equipment (e.g., hotplates or RIE chucks). Pick up wafers from the major flat whenever possible. When carrying wafers, it is advisable to hold your other hand under the wafer in case you drop it. Clean your tweezers regularly.

Gloves

Do not touch the front or back surfaces of your wafers. Your gloves are always contaminated to some extent. Brand-new gloves can have plasticizer residues, or you may have accidentally touched something dirty or dusty. Unless you have a very specific reason, you should never touch your wafers with your hands. However, for some processes, such as transferring wafers with freshly spun thick photoresist (e.g., SU-82100) to a hotplate, it may be appropriate to carefully transfer your wafer by lifting it gently from the edges with a doubled-gloved hand (never pinch with your thumb and index finger). Another example would be to avoid transferring tweezer marks to a hard mask if you plan to do anisotropic silicon wet etching. Again, change your outer gloves immediately if they become contaminated for any reason.

Cleanroom Dry Box Storage Space

Research groups and companies can request dry box storage for active wafers, photomasks, and necessary lab supplies. Additional dry box space is available for a monthly rental fee. Lab storage is intended for active materials, not archival storage. You are not permitted to store chemicals or other hazardous materials in your dry box (evaporation sources and crucibles are allowed). You cannot store any items that are incompatible with cleanroom protocols (e.g., cardboard or paper). If you need special chemical storage, please make a request to the WNF staff.

In addition to access from the cleanroom, you can access your storage dry box without gowning by going through the non-cleanroom office/work space in Room 135. Walk past the back-end process room, turn left, and follow that corridor to the end. To enter this maintenance and storage gray space, you must put blue shoe covers over your shoes and step on the tacky mats. Before touching your dry box or items inside, put on a pair of cleanroom gloves. You must clean any materials you bring into the cleanroom by wiping with IPA and cleanroom wipes located at the entrance to this gray space.

Exiting the Cleanroom

- When you exit the cleanroom, sit on the inner bench and remove your outer gloves and your white cleanroom boots. Leave your blue shoe covers in place while in the gowning room, and avoid putting your blue shoe-covered foot down in the cleanroom areas or your white boot down in the middle gowning area.
- In the middle gowning room area, remove your coverall first, and your hood second, leaving the hairnet, blue shoe covers, and inner gloves in place. While removing the coverall, do not allow the sleeves or upper part of the suit to touch the floor. Place your coverall on your hanger. Snap your hood to the collar snap of your cleanroom suit, with the outside facing outward and on the outside of the suit. Snap your boots to the legs of your coverall. Clip your ID clip to the suit. If you do not have an ID clip, ask a WNF staff member for a labeled clip.
- Continue to reuse the same gown upon each entry. On Monday evening or Tuesday morning, all gowns are sent out for laundering. The first time you enter the cleanroom after a garment change, get a new cleanroom suit and locate your nametag.
- If your cleanroom garment becomes soiled, do not place it back on the coat rack where it can contaminate other garments. Instead, place the garment in the laundry bin and get a new suit on your next entry.
- You must swipe your keycard each time you exit the cleanroom. This is required so that there will be an accurate record of who must be accounted for in an emergency.

Cleaning Up After Use

As with any shared space, you must thoroughly clean up every space you have used immediately after use. It is not okay to stage or store your project in a fume hood, on a workbench, or on a piece of equipment, as someone else has likely reserved time there after you. When they arrive, the space must be exactly as you found it—

all chemicals, beakers, and tools put away, all surfaces wiped down, everything powered off and reset to an open configuration, and you logged out.

If your project requires staging or will take multiple days to complete and cannot be stored in your dry box, it is important to arrange with the facility manager for an appropriate space to keep your project that will not interfere with others' use.

Chapter 4
MEMS Fabrication Process

As discussed in the Introduction, MEMS is an iterative process, repeating the same steps time and time again until the final design is achieved. Below is a graphic that illustrates this basic process. You start with a silicon wafer and apply a photosensitive coating to it that is called photoresist. When you expose your coated wafer to UV light, the regions blocked by your glass mask (containing your pattern for that layer) are not exposed. Thus, those regions are not resistant to the chemicals and will be washed away when rinsed. Then, when you expose your wafer to wet or dry etching, these same regions are vulnerable to the etching process and get etched away. The depth and shape of the etching is dependent on the process used, the amount of time, and the crystal orientation of the wafer. When completed, the wafer is rinsed in a different solution, such as acetone, and the final result (for that layer) can be seen.

Then you begin the process all over again and spin coat on another thin layer of photoresist and repeat the cycles. The etching process may be different, you may insert "sacrificial layers" that are thicker than photoresist that can be removed later, or you may deposit material rather than remove it, but the iterative cycle is essentially the same through to the finished design (Fig. 4.1).

This chapter is devoted to describing each of these steps in more detail and discussing some of the options and processes available for the steps.

Establishing a Process Flow Sheet

As is true for all manufacturing processes, you will need a process flow sheet for microfabrication in the cleanroom. Unlike typical flow sheets, however, this one is going to require some trial and error. You will not only need to decide upon all the processes you need to do and in what order, but you will need to determine the thickness of each film, the time, intensity, speed, composition, and equipment settings for every single process.

© Springer Nature Switzerland AG 2019
D. Munro, *DIY MEMS*, https://doi.org/10.1007/978-3-030-33073-6_4

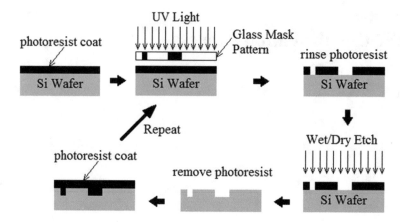

Fig. 4.1 MEMS iterative fabrication process

This is why you need to bring in your laptop and why you will want to have a cleanroom notebook. No matter how well trained you have been on a piece of equipment, every MEMS design is unique and every wafer has its unique properties. What worked on the last design with a different type of wafer is unlikely to work the next time. Everything becomes, "It depends." So, you will want to buy a full cassette of wafers in order to have some to practice with while you dial in the parameters that work for your design needs.

I recommend using a spreadsheet so that you can have several columns to record important values, such as time, thickness, or power settings. I also found it helpful to have a physical notebook to record values "on the fly," as most of the time, it is not convenient to have a laptop near the equipment. But I also record what I call "other considerations." If you were to read my notebook, you would find pages and pages of supplemental information about quirks of particular pieces of equipment, ways to "warm up" a machine or process, how to hold or rotate a wafer to achieve the best result, etc. A day in the cleanroom is overflowing with these details, and often weeks can go by before I return to a process flow step, so my notes are what I use to refresh my memory.

So, before you gown up and step into the cleanroom, have a process flow sheet prepared. Know in advance what variables you are going to have to define, and ask someone what a typical starting value or range you should expect, and start there. Get advice on how changing each value impacts your result, and get an idea of how sensitive the result is to each parameter. If it makes little difference if you etch for 1 minute or 10 minutes, but it makes a huge difference if your spin coat is one micron or two microns, then be aware of that and plan accordingly.

Deposition Process

One of the basic steps in MEMS processing is deposition of thin films of material. They must be uniform in thickness for best results, so either a spinning distribution process is used or some kind of vapor cloud is made that falls onto the wafer. These thin films typically have a thickness between one or two microns and 100 µm. Deposition is done by either a chemical reaction or a physical reaction.

Chemical reactions include processes such as chemical vapor deposition (CVD), electrodeposition, epitaxy, or thermal oxidation. These processes can create solid films directly from chemical reactions using gas or liquid. Each will be discussed in more detail below.

Physical reactions include spin coating, physical vapor deposition (PVD), and interpreted loosely, casting. For physical reactions, the material is deposited physically onto the substrate, and there is no chemical reaction which forms the material on the substrate.

Chemical Reactions

Chemical Vapor Deposition (CVD)

For this process, you place the wafer inside a reactor and supply a number of possible gases. This causes a chemical reaction to occur in a vapor cloud, which condenses on all the surfaces inside the reactor. Any parts of the wafer that are not protected by photoresist or another removable film will be exposed to this condensation, and the material will solidify and adhere. The protected regions will also be coated, but when the film is rinsed away, the condensate will be removed as well.

The two most common CVD technologies for MEMS are low-pressure CVD (LPCVD) and plasma-enhanced CVD (PECVD). The LPCVD process requires very high temperatures (greater than 600 °C) and has a slow deposition rate, but it produces high-quality, uniform layers. The PECVD process can operate at lower temperatures (as low as 300 ° C), because the plasma provides extra energy to the gases; however, the film quality is not as good as LPCVD. In addition, LPCVD can only expose one side of a wafer at a time and is limited to between one and four wafers coated simultaneously, whereas PECVD systems expose both sides of 25 or more wafers at a time.

Electrodeposition

This process is often called "electroplating" and refers to coating electrically conductive materials. Silicon is a semiconductor, so it can be coated with a metallic film, such as gold, platinum, nickel, or copper. For medical devices, only a few metals are acceptable for biocompatibility, such as gold and tin, so be sure to discuss your material limitations with the technical staff before designing your process flow sheet.

For electroplating, the wafer or component is placed in an electrolytic solution and an electrical potential is applied. First, you dip the wafer or component into a strong acid to clean it. Then, there is a vat of liquid electrolyte, which usually contains a metallic salt of the material you are hoping to coat with (such as platinum). You connect the wafer or component to the cathode (−) electrical lead and the "source metal" to the anode (+) electrode, which is usually a noble metal (like a platinum bar) with the other electrical lead. When an electrical potential is then applied, a chemical "redox" reaction occurs. Metal ions from the platinum bar are eaten away and are attracted to the cathode, which is the item you want to coat. Because this is a chemical reaction, gases are released during the process. If you need to coat in gold, be aware that strong chemicals, like cyanide, are required to enable the process to work. Typically, coatings range in thickness from 10 to 30 μm, but the full range is about 1 to 100 μm.

There is a second electrodeposition process called "electroless plating" which relies upon the electrical potential in the solution and does not need an external power source. Deposition occurs spontaneously on any surface where there is a high electrochemical potential (as compared to the solution). As you might expect, this process does not produce as uniform or precise results, but since the wafer or component does not have to connected to an electrical source, it is very convenient and can produce good enough results for many applications, such as the need to protect the exterior of a housing from corrosion in the salty environment of the body.

Epitaxy

Epitaxy is an interesting phenomenon that reminds me of how silicon wafers are created. To make a silicon wafer, you start with a molten supply of highly pure silicon in a vacuum chamber. A single seed crystal (just one) is introduced on a pull rod. When it makes contact with the molten silicon, identical silicon crystals spontaneously form. The pull rod is both spun and slowly extracted upwards for 1–2 days. The result is a column of pure silicon several feet in length that can then be sliced into wafers.

The process for epitaxy (usually done in the vapor phase) is similar to CVD, but if the wafer is a semiconductor crystal (like silicon), it is possible using epitaxy to continue building on the substrate wafer and have the *same* crystallographic orientation with the wafer acting as the seed for the deposition.

The most common type of epitaxy in microfabrication is vapor phase epitaxy (VPE). In this process, different gases are introduced in an induction-heated reactor chamber. Only the wafer is heated and it typically must be at least 50% of the melting point of the material to be deposited.

Epitaxy provides high growth rate of material and large thicknesses (>100 μm). One common application is to create an electrical isolation for the back of a silicon wafer, a technology called SOI, which stands for silicon on insulator.

Thermal Oxidation

One of the most basic deposition technologies is thermal oxidation, which literally means oxidation of the wafer surface in an oxygen-rich atmosphere. The temperature of the reactor is raised to between 800 °C and 1100 °C to speed up the process. Oxygen diffuses into the surface of the silicon wafer and forms an oxide film of the wafer's material (silicon dioxide). As the film thickness increases, diffusion slows and eventually stops, because it is self-limiting. Note that because it is forming a film of silicon dioxide, some of the silicon is consumed in the creation of the film. This is also the only deposition technology which actually consumes some of the substrate as it proceeds.

Thermal oxidation is a convenient way to provide electrical insulation on semiconductors, so use it whenever you can. Often, silicon dioxide is an interim step for other processes later in the fabrication sequence.

Physical Reactions

Physical Vapor Deposition (PVD)

Physical vapor deposition, or PVD, uses a source material to create a vapor cloud of minute droplets that are deposited onto your wafer. The two most frequently used technologies are evaporation and sputtering.

PVD is used when you want to deposit metals, something you will want to do frequently to create electrical traces, solder pads, and electrical connections. It is far more common to use than CVD, because it is lower risk, cheaper, and easier to do. The resulting metal film quality is inferior to CVD, and thus the deposited metals will have higher resistivity, but for most applications, PVD is more than adequate. For medical devices, it is again important to consider biocompatibility, and some common PVD metals should never be used, such as aluminum and chromium, which have high biotoxicity.

Evaporation

For an evaporation process, you place your wafer inside a vacuum chamber along with a block of the source material, such as gold. The source material is then heated to boiling and begins to evaporate freely in the vacuum. As it condenses, it falls onto the surface of the wafer, creating a uniform film of metal.

There are two popular evaporation technologies, e-beam (electron beam) evaporation and resistive evaporation, each referring to the heating method. For e-beam, an electron beam is used to heat and evaporate the source metal, and for resistive, the source metal is placed in a tungsten bowl or "boat," and the boat is electrically heated with a high current to make the source material evaporate.

Sputtering

Sputtering is convenient for microfabrication because the material is released from the source at a much lower temperature. You place your wafer in a vacuum chamber that contains the source material (called a "target"). The chamber is then minimally pressurized with an inert gas, like argon. The gas is then ionized by a power source, and the released ions bombard the target, causing atoms of the source material to vaporize and then condense onto the surface of the wafer.

Casting

Casting is mainly used for attaching polymers (but also glass and other nonmetals) to a wafer. Hydro-polymers, for instance, are useful for sensing biological substances in the body, as they can be made to swell in the presence of a target substance. For casting, the material to be deposited is dissolved into liquid form in a solvent and then spin coated or sprayed onto the wafer. When the solvent is evaporated, a thin film of the desired material remains adhered to the wafer. This is, in fact, how photoresist polymers are applied to wafers for photolithography. Thicknesses can range all the way from a single monolayer of molecules 10–30 μm.

Photolithography

Lithography is a process that involves multiple steps, mainly pattern transfer, alignment, and exposure, each of which has its own intermediate steps to consider.

Pattern Transfer

Lithography uses the exposure to light (usually ultraviolet, called UV) to change the physical properties of a film on a wafer. The film is typically a polymer dissolved in organic solvents that is spin coated onto the wafer at one to two micrometers in thickness. The solvent evaporates when the wafer is briefly baked, drying the film (called photoresist) on the wafer's surface.

Please note that there are both positive <u>and</u> negative photoresists as well as positive and negative pattern feature masks. Positive photoresist is polymerized by exposure to light, making it *less resistant* to removal. Negative photoresist is *more resistant* upon exposure to light.

Exposure to the light is controlled by inserting a glass slide (called a mask) between the UV light source and the coated wafer. The mask has regions which block light with a solid black. There are both positive- and negative-style masks. A positive mask ("dark field") blocks light on everything but the features and a negative mask ("light field") blocks light only on the features.

Choice of mask really depends on the amount of light area each one would block. For instance, if you have just a few features and are using a positive photoresist (which weakens the photoresist when exposed to light), you would want to use a negative mask so that most of the glass slide would be clear, or light field. You could then see what you were doing more easily. If you accidentally chose a positive mask, the entire glass slide would be black except where the few features were located, and during alignment, you would need to hunt around under the microscope until your features appeared in the small clear windows. However, if you are patterning a large number of repeated, complex features with critical dimensions, then you may want a positive, or dark field, mask that defines the blank spaces in black. In general, choose your photoresist and mask using the same principles as for macroscale machining. You want to control for the important dimensions, not the unimportant ones.

Get advice from the technical staff about what combination of photoresist and mask will provide the best quality result for your critical design features. In general, it is easiest to stick with one type of photoresist and switch your masks back and forth, as the CAD software makes switching from positive to negative mask straightforward.

Not to confuse things further, but it also matters if you are etching or depositing material onto the wafer. Let us say that instead of etching your few small features, you are hoping to expose those regions to sputtering to deposit metal on them. It might then make more sense to switch your photoresist instead of your mask. Negative photoresist (strengthens the polymer where exposed to light) can be used with a negative mask that is mostly clear. This way, you would still be able to see what you are doing, and you would be able to expose only the small features to sputtering adhesion. The rest would be stripped away (Fig. 4.2).

Lithography is the principal method for microfabrication patterning, because it is quick and economical to do. It is used for multiple processes, including pattern transfer for etching and protection of regions for deposition. After an etching or deposition process, the photoresist is chemically removed (or "stripped") with acetone or other solvent. This process is often called "lifting off" the photoresist. Using photoresist films for etching is much more common than for deposition, as most deposition processes require high heat that would evaporate the photoresist. This is another reason sputtering is a popular deposition process, because it is compatible with photolithography.

Fig. 4.2 Positive and
negative masks with
negative photoresist

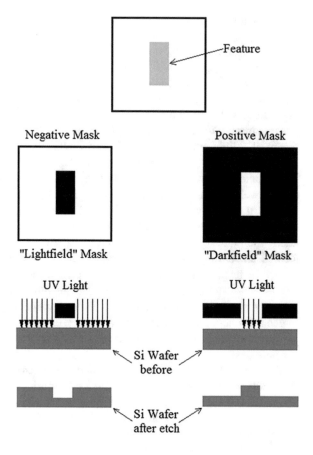

Alignment

Almost every MEMS device requires more than one layer to create. In order to build
up a multilayer device, you have to align the wafer and the masks precisely for each
lithography step. The first pattern transferred to a wafer usually includes a set of
alignment marks, which are high-precision features that are used for referencing
subsequent masks. Often, alignment marks are necessary in subsequent patterns, as
the original alignment marks have been obliterated or deformed by a processing
step. Thus, it is important for each alignment mark on the wafer to be labeled so it
may be identified and for each pattern to specify the alignment mark it uses (and its
location). By providing the location of the alignment mark, it is easy for the operator to locate the correct feature in a short time. Each pattern layer must have an
alignment mark that registers it correctly with the previous layer.

Silicon wafers come with flats on their perimeter, and these are used for gross
alignment. They define the x-y axes for the crystal's orientation. When placing your
wafer in the photolithography machine, use the flats to orient it correctly. The first

Fig. 4.3 Sample of crosshair-style alignment marks

Layer 1   When Aligned   Layer 2

pattern is typically aligned to the primary wafer flat (as there is no pattern on the wafer for the first pattern to align to). Subsequent patterns are aligned to the first pattern, using a microscope and the equipment's micrometer knobs to visually dial in your wafer's exact alignment to the alignment marks.

Usually, the alignment marks can be part of the mask and transferred to the wafer. A combination of lines and "crosshair"-style alignment marks works well, and many lithography machines have a slit designed for rough alignment of a line alignment mark. The crosshairs are typically paired with inverted crosshairs as shown in the figure. Sizes are usually around 50 μm square, with the lines 2–5 μm in width (Fig. 4.3).

There should be multiple copies of the alignment marks on the wafer. Etching processes can deform and obliterate alignment marks, so strategically place several around the wafer. I have found that having pairs of alignment marks well-spaced, but in horizontal and vertical alignment with each other, works well.

Please note that the equipment used to perform alignment may have limited travel and therefore only able to align to features located within a certain region on the wafer. Also, the geometry and size of your alignment marks may vary with the type of alignment you are performing, so the lithographic equipment and type of alignment to be used should be considered before locating alignment marks. At a minimum, it requires two marks (preferably spaced far apart) to precisely correct for rotation.

Exposure

The exposure parameters you will need to transfer the pattern to your prepared wafer depend primarily on the wavelength of light and the dose (time of exposure) required. Different photoresists have different sensitivities, but the dose required per unit volume of photoresist is somewhat constant. For UV light, the typical exposure is on the order of a few seconds at most. Physics effect your results, however. For instance, a highly reflective layer under the photoresist may result in a higher dose than if the underlying layer is absorptive. The dose will also vary with photoresist thickness.

At the edges of your pattern mask, light is scattered and diffracted, so if an image is overexposed, the dose received at the edges will increase. If you are using a positive photoresist, this will result in the photoresist image being eroded along the edges, causing a decrease in your feature sizes and a loss of sharpness at corners. When using a negative resist, the photoresist image is expanded, causing the features to be larger than desired, again accompanied by a loss of sharpness at corners. If an image is underexposed, the pattern may not be transferred at all, and in less severe cases the results will be the reverse of those described above.

If the surface being exposed is not flat, the high-resolution image of the mask may be out of precise focus for features that are nearer or further away. This is one of the limiting factors of using photolithography for MEMS. High aspect ratio features will also be problematic for obtaining even thicknesses of photoresist coatings, and a thick coating, as we have learned, also impacts exposure.

Summary of Photolithography Steps

Typically, photolithography is performed as part of a well-defined process, which includes the wafer surface preparation, photoresist deposition, alignment of the mask and wafer, exposure, and development. The standard steps found in a lithography sequence are the following: (1) dehydration bake, (2) HMDS prime, (3) resist spin coat, (4) soft bake, (5) alignment, (6) exposure, (7) postexposure bake,(8) develop, (9) hard bake, and (10) descum; however, not all lithography procedures will require all the process steps. I normally use only steps 3, 4, 5, 6, and 8.

1. Dehydration bake: Dehydrates the wafer to improve photoresist adhesion.
2. HMDS prime: Hexamethyldisilazane is an adhesive used to improve the adhesion of photoresist.
3. Resist spin coat: Coats the wafer uniformly with photoresist by spinning.
4. Soft bake: Removes some of the solvent in the photoresist, which reduces its thickness and makes it more viscous.
5. Alignment: Aligns the pattern on the mask to the features and/or alignment marks on the wafer.
6. Exposure: Exposes the coated wafer to UV light and creates the desired pattern.
7. Postexposure bake: Removes more of the solvent in the photoresist, which makes the photoresist more resistant to etching.
8. Develop: Removes the photoresist after exposure (exposed coating if resist is positive, unexposed coating if resist is negative). This is usually a wet process with solvents done in a bath.
9. Hard bake: Removes most of the remaining solvent from the photoresist.
10. Descum: Removes photoresist scum that may be clogging the pattern. It also helps you open up sharp corners.

As with all microfabrication processes, photolithography requires some trial and error. Avail yourself of the technical staff's assistance as needed to fine-tune a photolithography process that works well for your specific situation.

Etching

MEMS devices are typically micromachined using etching. In general, there are two etching processes: wet etching and dry etching. Wet etching is done in an immersion bath, where the silicon wafer is exposed to a chemical solution, such as potassium hydroxide (KOH), that removes material. Dry etching is done with an electron beam sputtering process or with a vapor cloud of reactive ions that remove material.

Wet Etching

Wet etching is also called "bulk micromachining" and is the oldest of the microfabrication technologies. It comes in two varieties: isotropic and anisotropic. Isotropic is supposed to etch equally in all directions, but quite often, the lateral direction (in the plane of the silicon wafer) etches much more slowly than the direction normal to the plane of the wafer, so the etchant solution needs to be vigorously stirred to both increase the etch rate and improve the lateral direction etching process. However, isotropic processes will cause undercutting of the mask layer by the same distance as the etch depth.

Anisotropic etching is dependent on the crystallographic orientation of the wafer. Crystal orientation is defined by Miller indices. These define the faces of the crystal and thus the planes at which the crystal will cleave and/or etch along. Miller indices use the (x, y, z) axes to define the perpendicular (or "normal") face of the crystal. Thus, (100) is the plane that is normal to the x axis, (110) is 45° to the x and y axes, and (111) is 45° from all three axes as shown below (Fig. 4.4).

For anisotropic wet etching, you must understand that etching proceeds at different rates depending on the crystal orientation. For example, if you have a (100) wafer and create a square opening feature on the surface of the wafer using a photoresist pattern, as you etch the (111) crystal planes will dictate the shape of the hole. Instead of perpendicular sidewalls as you etch deeper into the silicon, you will get an upside-down pyramid shape. There is no way to completely avoid this phenomenon, so one of the specifications you need to make on your wafer is the crystal orientation that is normal to the surface. If you want square sidewalls, try a (110) wafer.

Fig. 4.4 Miller index
plane orientations

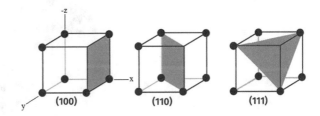

In spite of its limitations, wet etching is still a very good option for many MEMS devices, especially where the shape of the well, beam, or structure does not need to be square. It is a very cost-effective and rapid process which lends itself well to scaling up to mass production.

Dry Etching

With dry etching it is possible etch almost straight down without undercutting, which provides much higher resolution, but the process is both extremely slow and the necessary equipment is expensive. Dry etching comes in three varieties: reactive ion etching (RIE), sputter etching, and vapor phase etching.

For RIE, the wafer is placed inside a reactor and gases are introduced. Using a power source, the gas molecules are broken into high-energy ions. When the ions strike the surface of the wafer, they react and form a gaseous cloud. The ions at higher energy levels knock electrons off the surface and etch without a chemical reaction, so etching is a complex formula that must be balanced by settings on the machine. The main advantage of RIE is its ability to etch vertically, normal to the surface of the wafer. Since under-etching can still occur, the process is phased to introduce a protective coating on the sidewalls after each etching cycle. Under ultra-high magnification, you can see the ridges that are created by this cycling. The images below are from my MEMS device fabrication. At lower magnification, the 20-µm-wide fingers appear to have very sharp edges, but at higher magnification, you can see the ridges (Figs. 4.5 and 4.6).

The above etching was done using a subclass of RIE called deep RIE (DRIE), which has become a popular alternative, as you can achieve vertical sidewall etching for depths up to hundreds of microns. DRIE process capabilities are mostly limited by the width of channel and the access of the coating materials and the energized ions to reach the required depths. Typically, depths of 20–50 µm are much more typical.

The most common DRIE method is based on the "Bosch process," named after the German company Robert Bosch which filed the original patent. The cyclic appli-

Fig. 4.5 Topside etch of silicon wafer

Fig. 4.6 Same topside
etch at 10,000x
magnification

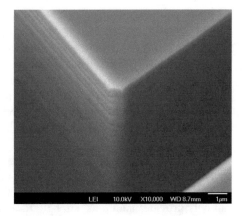

cation of a polymer coating was invented by them to protect the vertical sidewalls. The horizontal surfaces are also coated, but when bombarded by the high-energy ions, they are easily etched away. The polymer coating is only subjected to the chemical reaction, and it is resistant to it. As a result, etching aspect ratios of 50 to 1 can be achieved with etch rates that are 3–4 times higher than wet etching.

Another dry etching technology is called inductively coupled plasma (ICP) RIE. In this type of system, the plasma is generated by an RF-powered magnetic field. ICP can achieve very high plasma densities. A combination of parallel plate and inductively coupled plasma RIE is possible. In this type of system, the ICP is employed as a high-density source of ions to increase the etch rate. A separate RF bias is applied to the silicon wafer to create directional electric fields to obtain more anisotropic etch profiles.

Alternatives to RIE include sputter etching and vapor phase etching. Sputter etching only uses ion bombardment and has no chemical reaction. Vapor phase "dry" etching only uses a vapor cloud of chemicals and is similar to wet etching, but done in the gaseous state. The two most common chemicals are silicon dioxide etching using hydrogen fluoride (HF) and silicon etching using xenon diflouride (XeF2). Since it is a chemical reaction, you need to be careful that by-products of the reaction do not condense on the surface and interfere with the etching reaction.

When designing your MEMS device, it is important to remember that dry etching is significantly more costly than wet etching, and often the resolution and verticality provided by dry etching is not justified by the design specifications.

Releasing

After performing multiple microfabrication processes on your silicon wafer, you will need to release the microscopic MEMS device in order to use it. Although you may have created hundreds of instances of your sensor or actuator on the wafer, not all of them will be perfectly formed, and many will get damaged during release, reducing your net yield—a yield of 50% is considered acceptable.

There are three common techniques to release your device, depending on the fragility of the final structure. If your sensor is fairly robust and does not have fine, delicate structures or tight tolerances between sensors, you can dice the wafer. This is done with a very thin saw blade and is a rapid process done outside of the cleanroom. Much like a miniature table saw, the wafer (or saw arm, depending on the machine's design) is slid across the wafer in straight lines, creating a bunch of strips. The strips are turned 90 degrees, and the individual sensors are diced off. To protect the wafer from debris or fracture, dicing tape can be adhered to the surface prior to dicing.

A second technique is to initially adhere your wafer to a backer wafer using photoresist. Since wafers are extremely flat and smooth, the surface tension of the microlayer of photoresist securely affixes the two wafers together. A backer wafer not only provides additional strength for handling, but it also helps protect equipment from inadvertent damage, especially when etching through the entire thickness of your wafer. After final processing, the backer wafer is soaked off using an acetone bath, and the sensors are released as individual components into the bath. Each sensor is then retrieved using tweezers.

A third technique, and the one that I used, is good for MEMS devices that are delicate and need to maintain precise alignment. For this technique, the component is held into the wafer by small "tethers," but over 90% of the sensor is free floating with an etched "moat" around it. To release my sensors, I adhered a small strip of dicing tape over the top and then manually broke the tethers using a needle while looking through a microscope. I could then lift the sensor out of the wafer using the tape and glue it to my specimen. In my particular case, my sensor had two "halves" that were temporarily held in alignment using additional tethers until after mounting. These were also manually released from each other by manually breaking the tethers using a needle while looking through a microscope. I could have also separated the halves using an electron beam laser or etcher.

Mounting

Your sensor or actuator can be mounted to its circuit or package using one of several techniques. One common method is wire bonding. The device is glued in place on your specimen and then electrically connected to the rest of the system with gold wires that are a few microns in diameter. This technology is reminiscent of a sewing machining. There is a spool of gold wire that is threaded through a needle nozzle. The base tray holding your specimen moves back and forth as needed to dispense some wire, melt and fuse it to the first solder pad, dispense the needed length of wire to move to the second solder pad, and then fuse and melt the other end of the wire. As the wires are tiny and fragile, you typically do this 3 to 5 times to ensure you have a strong electrical signal pathway.

A more robust option for mounting your device is to solder it with thicker, insulated wires. This can be done by hand using a biocompatible, lead-free (preferably fluxless) solder under a microscope, which is challenging. To avoid damaging the delicate, microfabricated sensor, you can solder from a larger solder pad to the rest of the circuit and wire bond between the solder pad and the sensor.

A third option is to do surface-mount soldering technology (SMT). For this technique, the base of the sensor has solder pads sputtered onto it. A solder paste is applied to these pads (or to the circuit's matching solder pads), which temporarily holds the sensor in place due to surface tension. The circuit board is then heated, which melts the solder (a process called "reflow"). When it cools, the sensor is soldered to the circuit board.

When heating to reflow temperatures is not advised, a final option is thermal compression bonding. For this technique, a small bead of gold is adhered or patterned on each of the sensor's solder pads (usually on the bottom side). Matching solder pads on the circuit board are aligned, and the sensor is pressed into place under low heat. This deforms the gold bead and causes the gold molecules to join with the gold in the solder pads on each face, creating a permanent bond. The advantages of this method are: low heat, no solder flux or moisture trapped beneath the sensor or other components, so weak electrical conductive pathways or shorts from solder spreading around the board, no released contaminants, and no cleaning required.

Integration and Packaging

The unexpected truth about MEMS device design is that microfabricating your device is the easiest part. The process flow is relatively straightforward, and the yield of good devices is high. Mounting the sensor to the rest of the circuit is significantly harder than fabricating the sensor, as you can understand from reading the prior section, but by far the most challenging part is packaging. As a rule of thumb, packaging is about 75% of the design and fabrication effort for a MEMS device.

Somehow, you have to electrically integrate your MEMS sensor into a circuit, which was fabricated using different IC technologies that require power and are not typically used for reading a chemical or mechanical signal input from a MEMS sensor. Due to the tiny scale, the electrical impedance and noise are high, the sensor's output is minute, and the relative size differentials are huge. Even a tiny 0402 surface-mount component, like a resistor, is 4×2 mm, and the sensor will frequently have sizes one hundred times smaller than that at .04 mm (40 µm). So, you will need a paradigm shift in your way of thinking.

If you have microfabricated the electric circuit as well as the MEMS sensor, be cognizant of how you intend to mount your device and electrically connect it to other system components. If you have tiny sputtered wire traces between surface-mounted components, you may still need these to terminate in larger, spaced solder pads. Below is an illustration of an amplifier I used on my sensor. The amplifier is

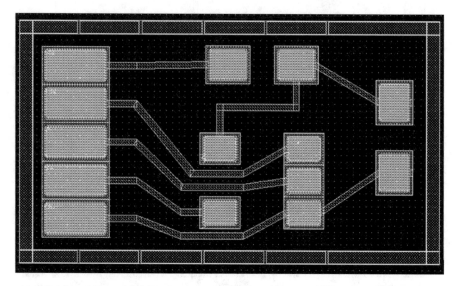

Fig. 4.7 Microfabricated circuit board with solder pads

called a low-Z amplifier and consists of a resistor, a capacitor, and an op-amp (operational amplifier) that are surface mount soldered onto the solder pads. The printed traces between the components connect the components into a simple circuit, but this then needs to be electrically connected to the rest of the system, which is done with a ribbon cable soldered to the five large solder pads on the left (at 1-mm spacing) (Fig. 4.7).

After integrating your MEMS sensor with your circuitry, you will need to package it in a protective housing. The housing may serve many purposes—electrical and/or temperature isolation, contaminant/dust protection, waterproofing, structural support, mounting sites, etc. The IC industry uses a lot of conformal coatings and lead carrier housings, which isolate things like a computer chip on a larger circuit board. For MEMS devices, where interaction with the environment is required—mechanically, physically, electrically, and/or chemically—the housing is more complex. It has to protect the component and yet still allow it to function for its intended purpose. For instance, MEMS accelerometers use a thin silicon beam across a well with a proof mass in the center. As the proof mass is accelerated upwards or downwards, the beams flex and emit an electrical signal that is amplified, converted into a digital voltage, and correlated to the relative acceleration. The housing needs to protect this delicate beam/mass system but still allow it to move freely. It also needs to prevent moisture from entering, which could get trapped beneath the mass, keeping it from moving. It further needs to keep the accelerometer cool, as silicon (like most materials) expands when heated, and this would throw off the calibration (Fig. 4.8).

Fig. 4.8 Conceptual
MEMS accelerometer in
cross section

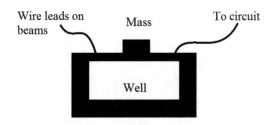

In conclusion, do not leave your MEMS component integration and packaging design as a problem to solve "later." It needs to be part of your design plan from the beginning, as your packaging requirements will dictate how you fabricate your device.

Chapter 5
Equipment

All of the NNCI facilities have a vast quantity of specialized equipment available for your use. The equipment is obviously expensive and must be operated correctly to avoid damage, and there can be multiple users waiting for their turn on a piece of equipment at any given time. Thus, common sense equipment use policies have been enacted by the facilities. The following is taken from WNF's users' manual.

Equipment Policies

- You must be trained and officially qualified before using any equipment.
- Do not use equipment for unapproved purposes.
- Tool owners must authorize all new processes.
- Your reservation will be void if you have not enabled the tool within 15 minutes of your start time.
- You must report equipment problems or damage in the online system.
- Do not use equipment that has been enabled by other lab users.
- Disable equipment when you are finished.
- You must submit a Billing Adjustment Request every time you need staff to adjust your billing. If you require an adjustment because you forgot to disable an instrument, you will be charged a fee.

Equipment can be divided into several broad categories: preparatory, masking and patterning, lithography, etching, building, releasing, mounting, and integration and packaging. The following serves as a partial list of the many thousands of equipment types found on NNCI's website.

© Springer Nature Switzerland AG 2019
D. Munro, *DIY MEMS*, https://doi.org/10.1007/978-3-030-33073-6_5

Preparatory

Cleaning

Wet bench—an enclosed fume hood with equipment, chemicals, and baths for various processes.

Spin rinse drying—a washing machine that holds a cartridge of wafers. It both washes and spin-dries the wafers.

Critical point drying—an alternative way of drying wafers that vaporizes the liquids.

Plasma/stripper—a machine to rapidly clean a wafer and remove whatever is on the surface.

Thin-Film Processing

Spin coating—usually a pair of spin machines with an array of liquids. The selected photoresist is poured on the wafer and then spun to evenly distribute. A second spin machine provides the baking/hardening.

Bonding—one of several machines used to bond wafers together.

Resist processing—a type of cleaning station that removes the developed photoresist, usually through a series of cleaner and cleaner baths.

Spray coating—an alternative to spin coating for applying a film to a wafer.

Dry Resist

A new option for applying photoresist is a method called dry resist. It is thin sheets of photoresist that are placed onto the wafer and then baked/melted into place. This provides a much faster and more uniform photoresist.

Masking and Patterning

Mask Making

It is a group of equipment used for making pattern masks. These interface with the CAD software to create black patterns on glass plates for the photolithography machine.

Contact Patterning

It is an alternative method for creating patterns on wafers which is often used with soft materials, like polymers. The pattern is transferred to the wafer via physical contact.

Inkjet

It is an alternative method for creating patterns on wafers using inkjet printing technology. Like contact patterning, the pattern is created using physical contact.

Laser

Laser can be used instead of a glass pattern and high-intensity light to pattern the photoresist or film on the wafer.

3D Printing

This is a relatively new technology to build up 3D structures on a wafer using 3D printing. Like other types of additive manufacturing, the printer can dispense more than one type of material—such as a structural material and a sacrificial/soluble material. After printing, the sacrificial material is rinsed away, leaving a 3D structure.

Lithography

Photolithography

It is a technique for patterning photoresist on wafers using a light source and a pattern mask.

Soft Lithography

Another term for contact patterning and/or casting, soft lithography is used for applying soft materials, such as polymers, biologics, and gels, to a wafer.

Electron Beam Lithography (EBL)

Instead of photolithography, the photoresist or film is pattern with an e-beam laser, producing a high-resolution and fine pitch pattern.

Etching

Chemical (Wet)

It is a bulk micromachining process that is done in a fume hood with chemicals, such as potassium hydroxide (KOH), to etch wafers rapidly. Usually, this is done anisotropically and the result is self-limiting along crystal planes.

Reactive Ion

It is a type of etching done with high-energy electrons bombarding and reacting with the surface of the wafer to etch precise, deep features.

Deep Oxide

It is another type of dry etching that relies upon chemical oxide reactions to remove material instead of ions.

Drilling

It is a microscopic drilling process used to create holes and vias through wafers.

Plasma

It is a type of electrical energy that provides high-power cutting and etching. It is very rapid but lower resolution than reactive ion etching.

Building

Sputtering

It is a machine that produces a cloud of a desired material, such as a metal, and allows it to fall and condense on the surface of a wafer.

Chemical Vapor Deposition (CVD)

It is a machine that chemically reacts gases to produce a cloud of a desired material, such as silicon dioxide, and allows it to fall and condense on the surface of a wafer.

Sacrificial Layers

These are materials, such as silicon dioxide, that support structures on the wafer until after fully processed. Then, the sacrificial layer is removed, often with wet etching, leaving free-standing structures.

Insulators

These are material coatings, such as glass, that are applied to a wafer to electrically isolate or insulate them from their environment.

Bonding Wafers

These are sacrificial wafers (also called backer wafers) bonded to the back of the processed wafer to provide extra strength and/or to protect equipment from processing steps, such as when etching completely through a wafer.

Releasing

Tethers

It is a way of holding devices in place during processing that will later allow them to be easily removed at low force.

Dicing

It is a saw for cutting wafers in order to remove individual devices.

Mounting

Soldering

It is a surface-mount or solder iron soldering to secure components and fabricated devices to a specimen, package, or electronic circuit.

Compression Bonding

It is a metal, such as gold, applied as a bead to an electrical contact point. It is compressed and fused to the opposite contact point.

Wire Bonding

It is a metal wire, usually gold, that connects one electrical contact point to another.

Adhesives

These are various glues, such as cyanoacrylate (CA or "super glue"), that are used to secure a device or electrical wire in place.

Integration and Packaging

Chip-Level Packaging

This refers to placing the fabricated device into a "chip," which is usually a plastic, rectangular housing with "feet" for soldering to a circuit board.

MEMS Fabricated Housings

It is possible to create microfabricated housings for a device that contain circuitry traces, antennae, soldering pads, compression bonding locations, and more. The device may be fabricated completely separately and later integrated into a microfabricated housing.

Chapter 6
Tools and Supplies

In order to begin fabricating MEMS devices, you will need some wafers and a number of personal tools for handling them. The wafers are extremely fragile, so it is necessary to use specialized tools, holders, and other supplies in order to not damage them.

Purchasing Wafers and Other Supplies

Characteristics of Wafers

Diameter—most wafers for MEMS are 100 mm in diameter (4 inches), but some facilities can support 150-mm and/or 200-mm wafers as well.

Thickness—most silicon wafers are 600 μm thick, but thinner ones can be obtained, such as 300 μm.

Single or double sided—if you intend to do processes from both sides of the wafer, you will need a double-side polished wafer. Otherwise, only a single side of the wafer is polished.

Resistivity—the lower the resistivity, the better electrons (electricity) will flow through the semiconductor material. If what you are sensing depends on the material itself, choose a low resistivity. If you want to block electrical conductivity, choose a high resistivity.

Insulating backside coating—the backside of the wafer can come precoated in an electrically insulating material to isolate the devices from their environment.

Crystal orientation—most users choose (100) crystal orientations, but for anisotropic wet etching, a different crystal orientation may be desirable.

Alignment flats—cut flats on the outer edge of a wafer for alignment in equipment.

© Springer Nature Switzerland AG 2019
D. Munro, *DIY MEMS*, https://doi.org/10.1007/978-3-030-33073-6_6

Other Supplies

Special solders—many commonly used MEMS solders contain biotoxic materials, which will cause regulatory approval issues later. However, there are a variety of silver-based solders that can be obtained in both a paste and wire spool format.

Chip-level packages—small plastic housings with solder feet are available for purchase to contain your MEMS device.

Patterning masks—often, masks can be ordered online directly from your computer and CAD patterns. They are typically delivered within a few days to a week.

Tools for Handling Wafer

Toolbox—a plastic box, preferably with a hinged lid and thumb clasp, is ideal for holding your various supplies. It should be long enough for tweezers and graspers and wide enough to hold a round wafer dish for single wafers.

Graspers, tweezers, etc.—it is not recommended to touch a wafer with your hands, as hands have contaminants, can leave fingerprints, and can easily break wafers. It is very difficult to pick up a wafer from inside of an equipment's tool tray, too, so you will want a small variety of graspers and tweezers to handle your wafer.

Wafer storage boxes—usually, wafers are purchased in quantity (from 10–25) and come in a storage cassette (box) with plastic separators between wafers.

Wafer individual dishes with lids—once processing has begun on a wafer, it is held in an individual dish with a lid. There is often what is called a "spider" in the dish to keep the wafer secured against the bottom of the dish when the lid is on.

Ways for Recording Data

Laptop or tablet computer—it is frequently necessary to have a computer in the cleanroom to record process or setting data for a wafer.

Camera—one of the quickest and easiest ways to record results is with a camera (no flash allowed). This is especially helpful for capturing visual information, such as images projected onto a screen that are not recordable to a computer.

Screen capture—when recording to a computer is feasible, the quickest way to obtain the image you want is with a screen capture. Then, save the file to your individual account folder, as the lab computers are erased frequently.

Lab computer into file—all computer files should be saved immediately to your individual account folder, as the lab computers are erased frequently. Also, subsequent runs of a process often overwrite the results of the prior process and those results cannot be retrieved at a later time.

Cleanroom notebook—a cleanroom notebook (which creates no particulate debris) is a great place to jot notes about equipment settings or special instructions. Use with an ink pen.

Chemicals and Special Use Materials

Hazardous material cabinets—all liquid chemicals are stored in cabinets, with flammables in their own fireproof cabinet.

Usage rules for materials, biologics, and more—if you have special materials, such as new chemicals or biologics, they must be approved before bringing into the lab and they must be well-labeled and stored in the designated cabinets.

From WNF's user manual, the following information about hazards was excerpted:

Understanding Hazards

Do not use or handle any chemical until you read and understand its label and safety datasheet (SDS). Understand the hazards, handling, storage, disposal, and emergency procedures for every chemical you use. SDSs are located at or below the Right-to-Know Workstation at the southeast side of the gowning room and are also available online. You also need to know evacuation routes and locations of eye-washes and shower stations.

General Safe Practices

- Do not taste, touch, or smell any chemicals.
- Do not mix, heat, dispose, or otherwise use chemicals in an unauthorized manner.
- Work with chemicals in an exhausted fume hood or wet bench.
- Use chemicals only in wet benches where they are approved.
- Never mix acids and solvents.
- Never dispose of solvents down water drains or water down solvent drains.
- Change your gloves if they might be contaminated.
- Label your chemicals.
- Do not place or store chemicals above the level of the wet bench surface.
- Never remove chemicals from the lab without permission.
- Use chemicals and cleanroom wipes sparingly.
- Do not interrupt users working with chemicals.
- If you are unsure of handling or safety procedures, ask questions.

Buddy System

Most chemicals used for cleaning and etching wafers are very dangerous, so it is required that another authorized cleanroom user accompany you while you are working at the wet benches. A buddy is required for all wet processing performed in wet benches on the marked side of the photolithography room. All dry processes are permitted without a buddy; however, it is recommended that you coordinate lab activities to ensure that at least one other person is in the vicinity. You may not assume that someone is your buddy if they happen to be in the lab. You must explicitly notify them that you need a buddy, and they must accept that responsibility. Your buddy may not leave until chemical operations are completed and you have cleaned up.

Chemical Classes and Storage

Acids

Acids are substances that donate protons when dissolved in water. Acids are used for etching metal and cleaning wafers, are generally corrosive, and can be toxic or water reactive (e.g., sulfuric acid). Acids are stored in the blue corrosives cabinets and in marked fume hoods. Metal etchants are also stored in marked fume hoods.

Bases

Bases accept protons and can increase the hydroxide ion concentration when dissolved in water. Many photoresist developers are dilute bases, and some concentrated bases can be used to etch silicon. Bases are stored in the upper gray cabinets and in marked fume hoods.

Oxidizers

Oxidizers are agents that are easily reduced and generally supply oxygen to chemical reactions. Examples in the lab include hydrogen peroxide and nitric acid. Oxidizers can react violently with organic chemicals.

Solvents

Although the term "solvent" refers to any liquid used to dissolve another material, in a cleanroom setting "solvents" are typically organic liquids that are flammable or combustible. We use acetone, isopropyl alcohol, methanol, n-methyl pyrrolidone, dimethyl sulfoxide, and a variety of others. Photoresists are usually photoactive polymers suspended in organic solvents such as propylene glycol monomethyl ether acetate (PGMEA) or cyclohexanone. Solvents and photoresists are stored in the yellow flammables cabinets on the east wall of the photolithography room. Photoresist strippers and solvent waste containers are kept in the blue cabinet next to the southeast emergency exit of the photolithography room.

New Materials Requests

Before bringing a new chemical into the cleanroom, you must submit a New Materials Request Form, an SDS, and a Standard Operating Procedure to the WNF website, which will be sent to the lab manager and the lab safety manager for approval. We do not permit long-term storage of any personal chemicals in the facility or wet benches without explicit permission.

Wet Bench Types

The photolithography room has metal (stainless steel) and plastic (chlorinated polyvinylchloride) wet benches. Organic solvents such as acetone, isopropanol, and SU-8 developer are not allowed on the plastic benches because they will dissolve the working surfaces. Acids and bases are not allowed on metal benches. More detailed restrictions are available in the documentation for individual benches that can be found online. You are responsible for understanding the specific requirements and chemical restrictions for each bench.

Avoiding Fumes

Fume hoods are designed to limit your exposure to chemical fumes and are equipped with pressure gauges and sash height sensors to ensure safety and proper operation. Although the benches are designed to turn off during exhaust outages, check the pressure sensors to make sure the pressure is within the acceptable limit. Many fumes in the lab are toxic, corrosive, or carcinogenic; hence, it is important to only work under the sash for very brief periods of time and only when absolutely necessary.

Personal Protective Equipment

You are required to use additional personal protective equipment (PPE) when working in the wet benches on the marked side of the photolithography room and when transferring chemicals to and from the corrosives cabinet.

Donning PPE

PPE consists of three items that should be donned in the following order: a chemical apron, a face shield, and chemical gloves. Check all items for damage before use. Look for cracks or pinholes in gloves, tears or holes in aprons, and scratches or cracks in face shields. If any gear is damaged (e.g., ripped apron or gloves), discard it and use another item. Rinse damaged items with deionized water and dry before disposing. Use care when putting on aprons to avoid ripping the seams, and make sure the apron sleeves are fully tucked under the chemical gloves.

Wearing PPE

Do not touch anything unnecessarily with the chemical gloves and treat them as though they were contaminated. For example, do not touch face shields, sashes, controllers, or any other equipment with the chemical gloves, and do not leave the photolithography room while wearing chemical gloves. It is acceptable to leave your chemical gloves on the edge of a wet bench while you work elsewhere. The apron must fully cover your shoulders at all times (i.e., make sure it is tied around the neck and do not let it slip off while you work).

Wearing PPE is not an excuse to act in an unsafe manner. Do not ever put your hands or fingers into a chemical bath, and always avoid splashing or spilling chemicals. Also, PPE provided by the WNF is only for temporary protection. It will not protect you from a spill, splash, or mist for a prolonged period of time.

Doffing PPE

Rinse and dry the chemical gloves, remove them, and hang them up. Hang face shields and avoid scratching the plastic. Lastly, remove the apron and be careful to avoid ripping it. If condensation has accumulated in the apron, use a wipe to dry the inside. Do not leave the apron inside out.

Labeling

Prior to filling, all chemical containers must be properly labeled even if you do not intend to walk away. You must include your name, the chemical name, and the date. If you plan to leave chemicals out after leaving the room, a phone number or email and an expected time of disposal must also be provided. Water must be labeled. If the chemical is not regularly used in the cleanroom (e.g., it was brought in after approval from the lab staff), list all hazards.

Pouring Chemicals

Assume that all chemical bottles are contaminated. Use a bottle carrier when transferring chemicals to and from storage locations. Immediately before pouring, always recheck the chemical label and make sure the chemical container you intend to use is set flat on the wet bench surface. Do not try to pour small volumes from gallon jugs; instead, transfer chemicals from gallon jugs to graduated cylinders or beakers, and then pour again from this secondary container. Use good judgment and do not overfill containers (i.e., do not fill them so close to the top that moving the container or disposing the chemical is unsafe). Never return poured chemicals to their original container.

Use containers that are compatible with your chemicals. For example, some chemicals or solutions, such as piranha (a mixture of sulfuric acid and hydrogen peroxide), cannot be stored in closed containers even for brief periods of time because it outgases and could cause an explosion. Also, hydrofluoric acid cannot be used with glassware because it will dissolve the container.

Chemical Bottle Cleanup

Use chemicals in partially used bottles before opening new bottles. Properly clean empty chemical bottles before disposal. Leave empty solvent bottles open in the hood in the designated location to evaporate. After the solvent residue has evaporated, fill the bottle half full with DI water and dump down a water drain. Repeat this process three times. Acid and base bottles must also be rinsed by filling the bottle half full with DI water, emptying the bottle into a water drain, and repeating at least three times. After rinsing, dry the outside of the bottle with wipes, use a black marker to cross out the label, and then write "Rinsed 3x" in at least two different locations on the bottle. Set the empty, rinsed, dry, and labeled bottle in the bottom shelf of the blue photoresist stripper cabinet.

Disposing Solvents

To dispose of used solvent, empty it into a solvent drain or into an appropriate waste container. A list of solvents allowed into the solvent drains is posted on the hood. Clean the chemical container with an acetone-soaked wipe, and then wipe thoroughly with isopropanol (IPA). Remove the label with acetone or IPA (do not bring a solvent squirt bottle into a plastic bench), rinse with DI at any of the plastic benches, and return the container to the drying rack.

Disposing Acids and Bases

Almost all acids and bases can be disposed into sinks in the plastic benches that drain to the neutralization system. Run the faucet so the solution will be diluted at least ten to one with DI water while carefully dumping the acid or base into the sink. Avoid splashing. Take the chemical container to the south side of the room and remove the label with acetone or IPA on a wipe and then return it to the drying rack.

Handling Small Spills

Attempt only to clean small spills for which your training and experience are appropriate, provided you can do so safely without taking unnecessary risks. Large spills or spills outside of wet benches should be treated as emergencies.

Clean small solvent spills with lint-free wipes and dispose them in the red solvent waste can. Then use acetone and IPA with wipes to clean the metal surface.

Clean acid or base spills in plastic benches by thoroughly rinsing the working surface with DI from a spray gun. Use care to avoid getting water in staff-maintained baths. Do not wipe up chemicals directly with cleanroom wipes without first rinsing and diluting the spill thoroughly. This is of particular concern with highly oxidizing agents (e.g., hydrogen peroxide or nitric acid), because of the potential fire hazard. After rinsing the surface, use a plastic scraper to move the water into the cracks between bench panels or into the sink (not into baths). Once you are sure that there is only water left on the surface, dry the remaining drops of water with a wipe to leave a clean, dry surface. Wipes are expensive; use sparingly.

Hotplate Safety

Hotplates are used extensively for baking photoresist and occasionally for heating solutions. Do not touch hotplate surfaces. Use extreme care when hotplates are used in proximity to flammable solvents or other liquids. Do not spill on hotplates or spray water on hotplates, and do not heat high vapor pressure solvents. For example, do not heat up acetone or isopropanol. It is acceptable to remove hotplates from a wet bench if you need more room or if you feel more comfortable working without one in the hood. If you need to heat an organic solvent or material in a bottle (e.g., SU-8), heat the container in a water bath, not directly on a hotplate.

Leaving Workspaces

After using a bench or other workspace, clean up all chemicals, chemical containers, wipes, and other materials (samples, tape, markers, notes, personal effects, etc.). Always leave wet bench surfaces clean and dry within comfortable arms reach and as organized as possible. It is not necessary to clean out the cascade rinse tanks or the very back of the bench tops.

Supply and Component Storage

Labeling—all items, including your wafer cassettes, toolbox, tools, wafer dishes, and other supplies, must be labeled with your name (at a minimum). Make a habit of providing full details on your wafers—process steps, process conditions, etc.—as all wafers look alike!

Cleanroom accessible storage—you may be assigned a dry box that contains your wafer cassettes, toolbox, and other cleanroom items. It allows quick access to items you are actively using.

Locker storage—other tools, such as personal measurement equipment, calculators, larger supplies, etc., can be stored in an assigned locker that is not accessible via the cleanroom. Items from locker storage can usually be transferred as needed to your cleanroom dry box without gowning up to enter the cleanroom (via a secondary entrance).

Chapter 7
Designing for MEMS

Additive Versus Subtractive Processes

MEMS devices are fabricated with one of two main techniques—additive or subtractive processes. The main difference is whether you are using the silicon wafer as the build platform or machining it to be the final device. Everything discussed in this book has been used for subtractive processes, as the equipment available in the NNCI facilities is mainly designed for subtractive techniques.

It is possible to do additive processes, too, but the design thinking is reversed. Instead of applying films in order to do selective etching, you apply films to build up areas. When the sacrificial areas are rinsed away, a proud structure remains above the surface of the wafer. In order to do overhangs, enclosed cavities, and other 3D features, a new planar surface must be created temporarily. This is usually some kind of filler glass, like silicon dioxide, that can be chemically removed later.

Thinking in Layers

Whether additive or subtractive, microfabrication is done layer by layer. A feature is etched and then covered with a film and a different feature is etched. Holes are etched to reach deeper layers on a third etch, and a metal sputtering is applied to create an electrical contact in the holes. Unneeded materials are stripped, and the process is repeated.

The wafer may be etched on both sides, too, so it is important to consider how the front and back side patterns will be aligned with one another. Alignment marks are necessary, but it is difficult to create a feature that goes through the entire thickness of the wafer, so you have to design in some overlap, where one etch is larger than needed in order to accommodate misalignment.

© Springer Nature Switzerland AG 2019

D. Munro, *DIY MEMS*, https://doi.org/10.1007/978-3-030-33073-6_7

Layout Software

Everything in MEMS design is microscopically small, so a CAD software is necessary to layout even the simplest of designs. Although there are many CAD software options for macroscopic design, they do not work well for microscopic, layered designs. Most MEMS designers use Layout Editor, which can be downloaded from the Internet. Universities typically buy a site license, and you can get access to it.

You will need to design in layers, thinking about one etching/build step at a time. All of the specifications discussed earlier in this book must be decided here—single- or double-sided wafer, positive or negative masks (per layer), positive or negative photoresists, position and design of alignment marks, etc.

Whether you are making the pattern masks yourself or having them made outside, you will need to output your CAD designs one layer at a time, specifying your requirements (i.e., positive or negative mask). The NNCI facility and/or the outside supplier will have a template for you to complete, which will ensure you have completely specified what you need.

Materials Allowed

Each NNCI facility has its own rules about what materials are allowed in its equipment. Many materials used in MEMS design are considered contaminants in IC chip manufacturing facilities and for good reason. If you are going to attempt fabrication of your MEMS sensor/actuator alongside your integrated circuit, get advice on what material limitations you will need.

MEMS sensors also sometimes use biologics or polymers, and these materials may not be allowed in some of the equipment, which will thus impact the sequence in which you conduct your processes.

Finally, it is important to be aware of the materials allowed for use in the human body. Often, you will need to make substitutions of materials in order to have a biocompatible device. Although it is possible to use biotoxic materials if they are hermetically sealed within a canister, such as titanium, realize this design choice will make your regulatory pathway more challenging. You must always prove that your device is safe and effective for the entire duration of its use. If your application is a device that remains in the body indefinitely, there is a high probability that a small percentage of sealed devices will develop a leak, allowing toxic materials to leach into the body.

Releasing from Wafer

Designing your MEMS device must include consideration for how you will separate your device from the wafer. The devices are tiny, and a typical wafer contains hundreds of copies of the device, so in your process plan, decide in advance what method you will use to release your individual sensor.

Dicing

One quick and effective way to release sensors from a wafer is to dice the wafer into strips and then cut each device off the strip. If this is your chosen method, you must provide straight, wide dead spaces between sensors so that the dicing sawblade can be used.

It is important to remember that dicing creates a lot of debris, a fine broken glass with smaller dust particles. You can damage your sensor by trying to clean off this debris. Therefore, consider using dicing tape, a thin film that can be adhered to the surface of the wafer. You will still need to be able to see the clear channels (blank spaces between devices) in order to cut with the tape in place. If your devices are not fragile, this allows you to cut up your wafer safely and then peel off the tape afterwards. Dicing tape also strengthens the wafer and reduces breakage.

A third option is to drop the diced devices into a bath of acetone or similar to rinse them off after dicing. You can pull out one sensor at a time for post processing, as the acetone will quickly evaporate without residue.

Moats and Tethers

Another releasing method is to use moats and tethers. Around each device, a moat is etched. The device is held in place by thin tethers at a few locations. These tethers can be thinned by etching the back side of the wafer, making them easier and more controllable to break.

Like dicing, breaking tethers will produce debris. To prevent this from damaging the device, you can use dicing tape to protect the surface and then break the tethers manually using a needle while looking through a microscope. Conveniently, the dicing tape will be adhered to your individual device, so you can use the tape to lift out the device and hold it in place in your packaging temporarily.

For very fragile devices, as well as for automating a process, you can cut the tethers using an electron beam laser. This has the added advantage of not requiring any tape. If this is your design choice, be sure to discuss with the technical staff the viability of using a laser to cut tethers, as there may be too much air or movement to allow loose devices to remain in place on the wafer during the process.

Dissolving Bond

Another approach to releasing devices from a wafer is to dissolve bonds. These can be photoresist bonds, such as with a backer wafer, or a sacrificial material that you dissolve away to release the devices into solution. No debris is generated, but the devices need to be rugged enough to withstand dissolution into a liquid and also retrieval with tweezers from the solution. If there are any soft materials used on the device, this method is not appropriate.

Integrating MEMS

You have created your microfabricated devices and released them from the wafer. Excellent! Now what? Somehow, you need to integrate them with the rest of your system. This usually means connecting them to some kind of electrical circuit that was fabricated separately.

Circuits and PCBs

Your MEMS sensor has to be integrated into a system, the main component of which is often an electrical circuit. Although the circuit can be fabricated via MEMS and laid out on a silicon wafer circuit board, it is much more common to have a printed circuit board (PCB) fabricated externally to which you can interface your device.

The PCB itself can be rigid, flexible, or a combination of rigid and flex (in which case, the PCB can be folded or coiled as needed to fit the form factor of the housing). Since everything on the PCB can be purchased off the shelf and soldered via traditional techniques (such as surface mount technology and reflow), there is really no value added to making your own PCBs. It can be done with bread boards and through boards for prototyping (especially handy for demonstration and testing purposes at a larger scale), but miniaturizing a circuit for a MEMS application is difficult and is better left to specialized subcontractors.

In most cases, the MEMS device is first mounted into a package, which is then soldered, wired, or bonded to the PCB. The package protects the device, provides a means of electrically connecting the sensor to the rest of the circuit, and can provide electrical isolation and/or moisture resistance as well.

Wire Bonding and Wire Leads

Wire bonding, unlike soldering, is done with a machine and does not require separately purchased wires/cables. The wire bonding machine is semiautomatic once you have selected the origin and insert/destination points for the wire. The machine fuses the first end to the origin, spools out the needed length of wire, and then fuses the wire at its destination. It is best for short distances (a few millimeters) and can only be used on electrical connections that have no stress or strain, as the wires are fragile and break easily.

Wire leads are small-diameter wires with an insulating coating that can be prepared and then soldered in place to connect the MEMS device to the circuit. This is usually done while looking through a microscope and can be tricky, as there is the soldering iron, the solder, and the wire to simultaneously hold and manipulate.

Soldering and Adhesives

Soldering can be done with a spool of solder wire or with a solder paste. Solder wire is convenient for manual techniques when wire leads and/or IC package feet need to be soldered. For surface-mount soldering, a small bit of paste is manually (or more typically automatically dispensed) onto each solder pad. The entire board is heated, and the solder melts or "reflows" to connect each device to its respective solder pads. Like solder wire, this can also be done component by component manually by using a soldering iron to reflow the solder, but this also heats the components.

Wafer Bonding

Usually, there is more vertical height available than is taken up by a single wafer MEMS device and its PCB. Thus, device systems can take advantage of vertical real estate by bonding wafers (or devices onto wafers) together in ways that establish an electrical connection at the same time. For instance, the backside of a MEMS housing can have a thermal compression bonding pad that interfaces with a corresponding electrical lead solder pad on a second wafer. The design possibilities are endless.

Packaging MEMS

Since MEMS devices are small and delicate, and they sometimes require a defined environment in which to operate properly, packaging is almost always required. This serves many purposes, as discussed previously, but a few of the key design considerations are addressed again below.

Insertion Forces

The amount of force you can apply varies, depending on whether you are applying it in compression, tension, or bending. Like macroscale devices, MEMS devices are strongest in compression, so you want your insertion force direction to be in that direction. The slightest tensile or bending force will often break a MEMS device, but firm compression is fine.

Materials

Silicon is a semiconductor, so it will conduct electricity, be susceptible to noise, and expand/contract with temperature just like other conductors. Often, the performance of the MEMS device depends upon electrical isolation of the sensor to reduce noise and eliminate stray conduction of an already weak signal. Be mindful of electrical noise, high impedance, and other issues when you are designing your device.

If you are intending to have wireless transmission of your output signal, remember that transduction depends upon many factors—the materials in use, thickness of those materials, shape of your sensor and its package/housing, impedance due to wire diameters and lengths, and more.

Waterproofing

One of the most challenging packaging aspects for a MEMS device is achieving a waterproof housing that still allows the sensor to perform as intended. The gold standard is a hermetically sealed housing, but this is expensive and it is highly unlikely you can do it yourself. Heat is required, and this can damage components or your device if it exceeds reasonable thresholds, so when designing, be sure to consult with vendors before finalizing your design.

Combining with Other Materials

Wafers can be made out of other materials besides silicon. Some of the more common choices include glass, quartz, and gallium arsenide. The wafer can also be etched to create a mold for soft, polymer, and cast MEMS devices.

Outsourcing of Processes

As has been noted already, some processes/parts will need to be outsourced in order to realize your design. Be sure to consider setup costs, minimum purchase quantities, lead time, and usability of your outsourced processes before committing. There are usually multiple sources for any given process, so shop around and get advice from the lab personnel on vendors they have used in the past and recommend.

Other NNCI Facilities

There are 16 NNCI facilities, and they all vary slightly from each other. Before going to an outside vendor, check what capabilities might be available to you through another facility. Since they are used to working with small businesses and academics on small batch MEMS devices, they are often your best resource.

Pattern Masks

One common step to outsource is creation of the pattern masks, as they require high precision, and there are companies that specialize in these. Creating a mask yourself is definitely feasible, but it can be faster and also better quality to have them fabricated elsewhere.

Electronics and PCB

Another common process to outsource is the electronics layout and PCB fabrication. There are companies that can create multilayer boards for you with the entire circuit layout optimized and tested. They can also place all your off-the-shelf components on the board and do the soldering of them, allowing you to focus on the science—sensing or actuating something with your MEMS device.

Packaging

Packaging design is at least 75% of the work in creating a MEMS system, so it makes sense to work with an outside vendor to design your packaging and have them make the packages for mounting your MEMS device (or even your whole system, including PCB).

It cannot be stated strongly enough to *design for how you intend to package.* A sensor is only as good and as useful as the package that contains it, so keep this reminder at the forefront of your thinking.

Testing

Design for testing. If you have a complex sensor or electrical circuit, make sure you can test the output at multiple places along the pathway for troubleshooting purposes. For example, include LEDs in your design so that you can quickly tell if it is receiving power. Have your software algorithm or the microcontroller output its status or provide ongoing feedback about what task it is performing. And have places along the circuit where you can query the output.

A lot of testing needs to be outsourced, as it is beyond the designer's capabilities to test. For instance, you may need accelerated fatigue testing or sterilization testing conducted. After the testing (or at intervals during the testing), you will want to measure parameters, which is where the ideas for troubleshooting discussed above come in handy.

Chapter 8
Soft Materials and Bioprinting

Soft Materials

A rapidly advancing area of MEMS is the fabrication of soft materials. These polymer-, paper-, and hydrogel-based devices have unique advantages as compared to traditional hard materials like silicon. They can be very low cost and fabricated without the need for expensive capital equipment by using molding techniques and bioprinting.

One common polymer material is polydimethylsiloxane (PDMS), also known as dimethicone, that belongs to a group of polymeric organosilicon compounds commonly referred to as silicones. PDMS is known for its excellent flow properties and is a hydrophobic viscoelastic material. It is also optically clear, inert, nontoxic, and nonflammable; thus, its applications range from contact lenses and medical devices to elastomers, and it is used in shampoos, food (as an antifoaming agent), caulking, lubricants, flexible electronics, and heat-resistant tiles.

PDMS is commonly used as a resin for soft lithography to make microfluidic devices. The process of soft lithography consists of creating an elastic stamp (using normal techniques of photolithography or electron-beam lithography) to transfer patterns of only a few nanometers in size onto glass, silicon, or polymer surfaces. The resolution, which depends on the mask, can be as small as six nanometers. It is possible to produce devices using this technique for optical telecommunications and biomedical research.

In biomedical MEMS applications, soft lithography is used extensively for microfluidic devices. Silicon wafers are etched to make channels, and PDMS is then poured over these wafers and left to harden. When removed, even the smallest of details are imprinted in the PDMS. Glass slides are used to cover the PDMS channels to create permanently sealed, waterproof channels. Researchers are using this technique to create lab-on-a-chip devices. PDMS is also being used to make synthetic gecko setae dry adhesive materials (Fig. 8.1).

© Springer Nature Switzerland AG 2019
D. Munro, *DIY MEMS*, https://doi.org/10.1007/978-3-030-33073-6_8

Fig. 8.1 Artificial gecko
setae on SEM at 500X
magnification

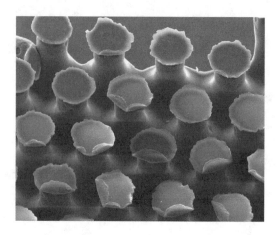

Bioprinting

Bioprinting is a form of 3D printing (also known as additive manufacturing) that utilizes hydrogels and other materials known as bioinks to create three-dimensional microscopic structures. Hydrogels are popular because they can be constructed with geometries that will hold cells or allow biological tissue to grow within them.

The advantage of using bioinks is their low cost. A desktop bioprinter can create a hydrogel scaffold for significantly less money than other MEMS processes, and the configuration of the geometry can be much more complex. Printing of bioinks is often combined with PDMS substrates to provide flexible, organic, and inorganic materials for use as sensors, tissue scaffolds, and chemical reactants.

Chapter 9
Imaging and Metrology

Optical Microscopes

There are a variety of optical microscopes in the NNCI facilities, including many inside the cleanroom. Because the MEMS devices are so tiny, you may need to check the quality and integrity of your devices after each process. Just remember that some process steps, like a spin coat of photoresist, are light sensitive. Some rooms have darkroom-style amber lighting, but not all, so double-check where it is safe to transport your coated wafer.

To measure the thickness of a photoresist film or check for complete coverage of a surface, you will need to use special light filters on your microscope; however, it is very easy to make mistakes and accidentally expose your film to stray light. In general, it is best to dial in the thickness parameters on a practice wafer and then do your imaging after the etching or deposition process.

SEM

Scanning electron microscopes are also available in the NNCI facilities. These machines have an imaging chamber that holds a cartridge, usually large enough for a 100-mm-diameter wafer and up to 10 mm of depth. For my research, I was able to place a 5.5-mm-diameter spinal rod into the cartridge without issue and scan a focused electron beam over a surface to create an image. An SEM uses a beam of electrons to interact with the sample, producing various signal reflections that can be used to obtain information about the surface topography and composition.

SEMs can obtain very high magnification images, such as 50,000x, but it is more typical to start at low magnification and slowly increase as needed. Obtaining a focused, high-resolution image requires skill and patience, but you will be trained, and assistance is readily available. You can capture images and use a separate soft-

© Springer Nature Switzerland AG 2019
D. Munro, *DIY MEMS*, https://doi.org/10.1007/978-3-030-33073-6_9

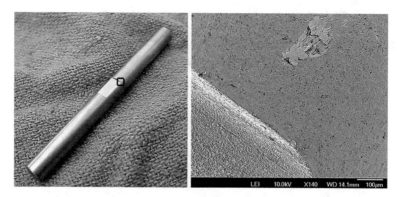

Fig. 9.1 Spinal rod (left) with 140× close-up of wire EDM cut (right)

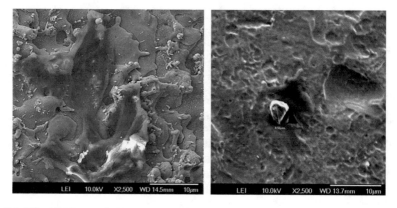

Fig. 9.2 Wire EDM cut at 2500× before and after electropolishing

ware to measure features of interest and/or make notes on the image. Like most imaging technologies, it is important to have a scale reference within the image field that you can use to calibrate the measurement tool.

Below are images of a wire EDM (electric discharge machining) cut I made on a titanium spinal rod. At low magnifications, the surface appears smooth, but at the microscopic scale of MEMS, it is unacceptably rough. I had the rod post processed with electropolishing to achieve a better surface finish (Figs. 9.1 and 9.2).

AFM

Sometimes, you need to physically measure the topography. If my goal had been to measure the surface roughness, I would have used an atomic force microscope (AFM). Interestingly, AFM used a MEMS fabricated needle, called a "cantilever,"

Fig. 9.3 AFM cantilever and tip

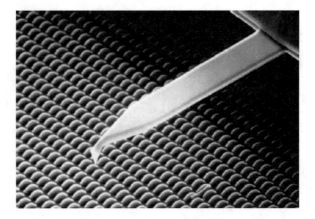

that has a small tip. The cantilever is both a force sensor and a force actuator. By pushing the cantilever across the sample, you can measure its topographic height that can be measured to nanometer accuracy (Fig. 9.3).

Image Storage and Retrieval

Images are the main outputs you rely upon while fabricating MEMS devices. Thus, it is critical to have a plan to capture and store them in an organized fashion. The NNCI facilities have servers where you can store your images, but be aware that it is usually an extra step to transfer the images from the local computer to the server where your personal folder is located. Although some systems have built-in image capture tools, many do not. You will need a strategy for capturing the images. You will find that a lot of the equipment does not allow you to rename the files as you (or has a very limited set of characters for names).

Some equipment provides a default name for your image file but will overwrite that file if you take another image; thus, you must consciously rename each file before proceeding. I also found that it was inconvenient to transfer files after each image, as there were steps involved that took me out of the main window. Thus, you may need to have a temporary naming scheme that you clean up later after transferring your files. For other equipment, there is no built-in image capture, so you have to use a screen capture or even a personal camera to get your images, so be prepared. My advice is to learn the limitations of the equipment and develop a naming strategy accordingly.

Also, get organized! Because of the nature of MEMS fabrication, you will generate a copious amount of images, and you will want a naming scheme and file folder scheme that allows you to quickly find the images you want to review or use. Record notes in your cleanroom notebook about your naming schemes and where you filed groups of images so that you can clean up your images at the end of each day after leaving the cleanroom.

Chapter 10
Testing

MEMS devices are complex, and there are many points in the process where something can go wrong. My advice is to start simple. Use everything at benchtop scale with electronics and components that are off the shelf until you are confident the MEMS sensor or actuator is performing correctly. Then, miniaturize or incorporate one thing at a time until you achieve your final configuration. This will save you a lot of time and frustration in the long run. The following sections are on types of testing you may need to ensure proper performance of your MEMS device, especially when you are down at the microscopic final size (Fig. 10.1).

Electronics

Signal Transduction

Is your device hardwired to a PCB and connectors, or is it using wireless communication? No matter which, your signal transduction is going to be a challenge. Pay attention to wire traces—are they wide enough? Thick enough? Fully formed with no nicks or narrow/thin spots? If you have solder connections, are they solid and clean? No stray/parasitic signal pathways? No fracture or erosion of the solder pads? If you have wire bonds, are they all fused at both ends? Do they accidentally touch each other or have enough length to cause electrical noise? Will they stay connected during expected use/handling conditions?

© Springer Nature Switzerland AG 2019
D. Munro, *DIY MEMS*, https://doi.org/10.1007/978-3-030-33073-6_10

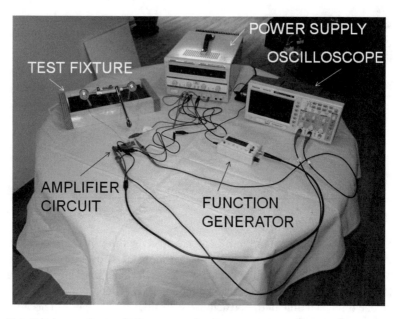

Fig. 10.1 Benchtop testing. MEMS device in fixture and amplifier PCB

Signal to Noise Ratio

The signals generated by your MEMS sensor will be tiny, so you need a means of amplifying them at the source of the signal, long before noise enters the equation. Everything is a potential source of noise at these microscopic scales—humidity, temperature, vibration, fluorescent lights, unshielded wires, nearby materials or power sources, even poor-quality microfabrication. It is important to design with signal quality in mind.

One thing you can do is have your MEMS device generate the largest possible output, such as moving your sensor's capacitive fingers closer together while also making them longer and deeper. Another thing you can do is convert to a digital signal as close to the MEMS device as possible and then amplify the digital signal. This way, it is easier to amplify just the signal without amplifying all the noise along with it. A third thing you can do is shield, shield, shield. Try to determine all your potential sources of noise and shield against their interference—encase your wire bonds in a plastic goop (to keep them from moving, touching, or interacting with each other), print metal, or an insulating glass on surfaces to prevent signal loss and/ or the outside environment from entering, and optimize shapes of your MEMS device and packaging to provide the cleanest signal pathway.

Data Storage and Transmission

How are you going to manage your data output? Does it "wake up" and respond when queried, store a value until queried, or continuously transmit data? Depending on your answer, you will need a specific data storage and transmission strategy. Whether you are using an implanted battery or temporarily charging a capacitor during signal transmission, you need to have a strategy for data storage.

Often, your frequency of data collection is extremely high (100 Hz, 1000 Hz, or more), which means you will be drowning in data output. It rarely makes sense to store all of that raw data, so you will need some means of consolidating it for storage and/or transmission. Let us say you are operating at 100 Hz and collect data for 3 seconds. That is 300 data points. Can those all be averaged into a single value? If so, great! It is much easier to store and transmit one value than 300. If the value is stored for later retrieval and transmission, you will need to store it digitally and have some kind of onboard storage, like a read-only memory (ROM) chip.

Analog-to-Digital Conversion

As mentioned previously, signals are easier to consolidate, amplify, and transmit if converted to digital. If you are making a cantilever and measuring the deflection of a proof mass, the electrical strain from bending the cantilever is an analog, continuously varying signal. Using an analog-to-digital converter (ADC), you can discretize the signal into millivolt steps. If you have an 8-bit ADC, then you can discretize your signal into 2^8 steps, or 256 increments. If you have 1 volt (1000 mV), then you will be able to detect a 1000/256, or 3.9-mV change. For a 12-bit ADC and 1000 mV, you will be able to detect a 0.24-mV change.

Batteries

Most implanted MEMS devices still require an internal power source in the form of a battery to provide enough power to operate. It is possible to use external excitation (such as from radiofrequency waves, often called RF) to charge up a device enough to transmit a small signal, and this technique is used for passive RFID tags, such as the chips put in pets to locate their owners. However, if your device needs to continually operate or has to transmit signals, you will most likely need a battery. Pacemakers use batteries that are hermetically sealed within a titanium canister, and they last 5 years without replacement. Although recharging batteries through the skin is also feasible, it is important to consider the risks that pose to the patient. For pacemakers, it is safer to perform a replacement surgery of the canister every 5 years, unplugging the leads into the heart and plugging them back into the new device.

Batteries are biotoxic, no matter what type is used, but some are safer than others. They come in all shapes and sizes, as well as various voltages, so selecting an OTS one for your needs is usually feasible. Because they are dangerous, you will need to pay careful attention to the waterproof sealing of your package and the intended lifespan of the implanted device when using batteries for power.

Finally, consider your power requirements. If you are purchasing OTS components, especially microcontrollers or amplifiers, know that they all have different peak power requirements. You will need to do a power profile for your system and determine the worst-case peak load. Let us say you add up all the power requirements of your components during worst-case usage and you need 5 mW (power in watts is volts multiplied by amps). That is not the whole story, as you also need to know the average power requirements and the amount of time you want your device to last. Usually, you want your medical device to last years, so the capacity of your battery must (1) meet your peak load needs and (2) have enough capacity to last, which is your average power multiplied by the amount of time, usually given in watt-hours (Wh).

Batteries are made of a wide variety of materials/chemistries, so not all batteries have the same performance. Lithium ion batteries have the highest energy density, so they are small and light, and they are commonly used in handheld electronics, like your phone. However, they have to have additional circuitry to control their charging and discharging rates, as well as monitor their temperature, as they can explode. This makes them undesirable for many medical applications. Lithium iron phosphate batteries are also compact and can deliver higher power, with the added advantage of not being susceptible to explosion or dangerous swelling. A third option, which is larger and heavier, is nickel metal hydride, which suffers from no memory effects and is safe, plus rechargeable, but these batteries self-discharge. So, you will have to find low discharge rate versions. For more information on batteries, go to http://BatteryUniversity.com.

Algorithms

You will need some means of converting (and perhaps digitizing, amplifying, or manipulating) your MEMS device's output into a usable form. This is accomplished via electronics and your circuit, but often you will need a microcontroller as well to manage signal traffic and control what happens in different phases of signal transduction. For prototyping, I recommend using an Arduino, which is a cheap, readily programmable microcontroller with thousands of software algorithms (in C or C++ programming language) that have already been written and made available for free. You can easily download a similar piece of code and tweak it to meet your needs. If you have never used Arduino, you can buy starter kits that include a guidebook with projects as well as all the electronic components you will need to build those projects.

After prototyping, however, you will need to miniaturize your microcontroller and program that device to meet your long-term needs. There are countless varieties of microcontrollers out there, and the programming languages are standard, usually C or C++, but sometimes others. If this is not your forte, there are many C programmers out there that can write you a suitable code and test it for you.

When dealing with software algorithms, the hardest part is troubleshooting. Do yourself a favor and incorporate visible output from your device, especially in the prototyping phases, but also in the final product. There is nothing more reassuring than an LED lighting up to let you know a device is working!

Mechanical

Structure of Device

You need to find a way to test the mechanical structure of your device. Your calculations may predict that 5 μm of bending at the base of your cantilever will produce 1 mV of output, but does it? You wanted a range of motion of 20 μm, but can the cantilever withstand that without breaking? Mechanical testing is hard to do at such low loads and small dimensions, so utilize your larger-scale prototype to do the majority of the work. It is relatively easy to make your MEMS device five times or ten times bigger in order to test it. Only after this step, use the AFM to apply the tiny loads to check your final device.

Remember that at these microscales, you have to consider other aspects of physics that you would normally ignore as negligible. For instance, cantilevered beams will bend under their own weight, and if they get close enough to the substrate, they can get "glued" to the substrate due to van der Waal's intermolecular attraction forces (called stiction, for "static friction"). Stiction in MEMS is when adhesive forces exceed restoring forces, so on this same cantilevered beam, stiction could be caused by moisture/capillary forces or even ionic forces. In addition to these examples, there are many other phenomena to consider, so get the advice of technical staff at the lab while designing your structure to understand what risks exist.

Mounting

Although robotic pick-and-place equipment, solder masks, and reflow soldering equipment exist for production runs of MEMS devices, you are going to start by doing everything manually. Releasing, handling, and mounting a tiny device are tricky. It requires care, patience, and an understanding of the ways in which the device is most vulnerable to damage. If you have long, cantilevered fingers, any force applied to those fingers will likely break them; however, in the surface plane, the device can be safely picked up with tweezers.

You are going to have to hand mount and solder your MEMS device, so keep that in mind and design "large" solder pads and/or places you can apply pressure. Think about how you are going to release the device from the wafer, mount it in place, and solder it, or you will suffer. I used a "moat and tether" design for prototypes. To remove a device from the wafer, I covered it with dicing tape and then broke the tethers with a needle while looking through a microscope. I then picked up the device with the dicing tape and adhered the tape on one side to my specimen. I folded the device back to reveal the underside and applied CA glue and then folded the device back down to adhere onto my specimen like you would a copper foil strain gauge. I applied light pressure with two needles in safe regions until dry and then gently pulled off the dicing tape, keeping the angle close to zero.

Then, I wire bonded from the solder pad "bars" to larger, more robust solder pads, and I soldered insulated wires between the solder pads and my PCB. I later trained a student to surface-mount solder all the OTS components to the integrated circuit, which required manually positioning the component on the dabs of solder paste and using a fine-tip soldering iron to reflow the solder just long enough to get it to melt, but not so long as to heat damage the component or the silicon wafer PCB. You will need to develop a procedure that works for your device.

Packaging

At some point, you will need to place your MEMS device and possibly its integrated circuit or PCB into a housing. This entire, complex process is simply called "packaging." I have separated out "mounting" as a pre-step, but that is often described under the umbrella of packaging as well. Be prepared for this to be difficult. IC chip manufacturers have it easy compared to MEMS fabricators. For a chip, you can often melt a conformal coating over the top and be done. With a sensor or an actuator, the device has to *interact* with its environment, and that makes packaging a challenge.

Also, if you are trying to wirelessly transmit your data, your packaging geometry, thickness, and material are prime considerations. Before you fabricate your packaging, I advise you to have an electromagnetic field (EMF) study done in a computational model with finite element analysis (FEA) and EMF computational fluid dynamics (CFD); otherwise, you will be guessing. Everything about your packaging will affect the quality of your signal transmission, so you will need to vary parameters until you find the configuration that works the best. Then after your device and package are constructed, you can have them tested in a chamber to make sure they are performing correctly. Both computational modeling and signal transmission testing can be outsourced to vendors who specialize in these fields.

Bonds

Wire bonding, usually gold wire of 18–60-μm diameters, is commonly used to electrically connect a microscopic MEMS device to the rest the circuit, but these bonds are fragile and prone to an array of problems. First, the bond connection can break, which happens frequently. Thus, you need to use three or more wire bonds for each connection. Second, the wires can break, so you need to protect them. One common way is to encase them in a polymer, such as parylene, to give them extra strength. Third, the wire is not insulated, so if two wires touch that should not, your circuit will not work correctly. In addition, "long" uninsulated wires can interfere with each other and increase noise in your signal. Encasing the wires in a polymer helps for this issue, too.

Fatigue

If your MEMS device works through repeated bending, vibration, thermal expansion, or other repetitive motion (or is subjected to these environments), you will need to do a fatigue analysis. For natural frequencies, this is also called modal analysis, and for controlled, repeated testing is called cycle life testing. Determine what your fatigue failure modes are using a standard decision tree, such as a failure mode and effects analysis (FMEA) and then design a way to test your device under each of those isolated and compounded conditions. FEA and a computational model are a good way to start, but eventually, you will need to do physical testing. You may need to fabricate other MEMS devices as test fixtures to then apply the conditions you need to do fatigue analysis on your sensor.

Soldering

Soldering while looking through a microscope requires a lot of fine motor skill and a steady hand. Many things can go wrong when you are soldering, such as overheating the solder or the component, wicking solder into regions it does not belong, breaking the MEMS device from pressure or heat, misalignment of components, accidentally "unsoldering" wires and connections, or losing your grip on something—the solder wire, the electrical wire, the component, or the integrated circuit board. I have made numerous fatal mistakes while soldering, so keep extra supplies and components on hand to account for attrition.

Environmental

Environmental testing is critical, especially for implanted or consumer use medical devices. Regulatory bodies require that you test for a host of environmental factors, assuming worst-case conditions and maximum end-of-lifetime points. If you are placing a sensor on/in a total hip implant, for instance, you will need to mimic *in vivo* conditions, such as 10 million loading cycles, a corrosive saline environment, electrical transmission through tissue (which can cause tissue heating), and worst-case scenario failure modes (i.e., the device leaks and the battery is exposed). The following are some common environmental considerations.

Waterproofing

One of my favorite sayings is, "Moisture finds a way." If there is a way for moisture to get into your packaging, it will. Maybe it will survive a few months or a year, but eventually, moisture will get in and destroy your device. Depending on what materials and components you have used and where the device is located, that moisture may allow biotoxic contaminants to enter the body.

So, your waterproofing strategy depends upon the risk. If the device is a long-term medical implant, you should choose a hermetically sealed package. Nothing else is going to gain regulatory approval, so incorporate those design constraints into your packaging from the beginning. If you are trying to transmit a wireless signal, you will need to have a radiolucent housing. If you need a large antenna, it may not fit in your packaging, and you will need to consider hardwired leads that connect to your MEMS device packaging through hermetically sealed leads. If you have a transcutaneous connector to an external device, such as an insulin pump, you will have to consider the additional hazard of infection and how that will impact your housing.

Sterilization

Medical implants need to be sterilized, but gamma radiation destroys electronics, and autoclaves are often too hot or too steamy for MEMS devices, which leaves ethylene oxide (EtO). This is safe and effective, but is not commonly available. You will not find it at the microfabrication facilities, but since most of the facilities are at research universities, you might find an EtO sterilizer at the hospital. Otherwise, to perform this test and how it impacts your device, you will need to send it to an outside vendor.

Wireless Signal Transmission

Although we often think of wireless as a single transmission frequency, that is entirely inaccurate. The FCC publishes a series of frequency bands that can be used for medical devices, and they range from kilohertz to hundreds of gigahertz. Frequency is the main parameter to consider for power requirements, distance for signal to travel, tissue heating, and compatibility with OTS receivers, so it is a key design feature to consider. In general, the higher the frequency, the smaller the antenna and the longer the distance you can travel. Unfortunately, the higher the frequency, the more tissue is heated and the more power is required, so it is a trad-eoff. Table 2 provides some commonly used frequency bands. Currently 13.56 MHz, which is in the RF band, and 2483.5 MHz, which is Bluetooth, are the most commonly used for medical devices (and electronics in general), but they may not be the most desirable bands for your application (Table 10.1).

Table 10.1 Common signal transmission frequencies for medical devices

Standard	Freauency	Data Rate	Range
Inductive Coupling	<1 MHz	1–30 kbps	<1 m
Wireless Medical Telemetry System	608–614 MHz, 1395–1400 MHz, 1427–1429.5 MHz	> 250 kbps	30–80 m
Medical Device Radiocommunication Service (MICS)	401–406 MHz	250 kbps	2–10 m
Medical Micropower Networks (MMNs)	413–419, 426–432, 438–444, 451–457 MHz		<1 m
Medical Body Area Networks (MBANs)	2360–2400 MHz	10 kpbs - 1 Mbps	<1 m
802.11a Wi-Fi	5 GHz	54 Mbps	120 m
802.11b Wi-Fi	2.4 GHz	11 Mbps	140 m
802.11 g Wi-Fi	2.4 GHz	54 Mbps	140 m
802.11n Wi-Fi	2.4–5 GHz	48 Mbps	250 m
802.15.1 Bluetooth Class I	2.4 GHz	3 Mbps	100 m
802.15.1 Bluetooth Class II	2.4 GHz	3 Mbps	10 m
802.15.4 Zigbee	868, 915 MHz, 2.4 GHz	40 kbps, 250 kbps	75 m
World Interoperability for Microwave Access (WiMAX)	2.5 GHz	70 Mbps fixed, 40 Mbps mobile	6–10 km
RFID/Near Field Communication (NFC)	13.56 MHz	1–464 kbps	<10 cm

Electrical Noise

When dealing with minute output signals, electrical noise has to be tested for, controlled, and filtered out. A lot of this can be done with good packaging design, but there are also clever electrical engineering solutions, such as band-pass filters. If your device operates at 100 Hz and transmits at 13.56 MHz, you can use a filter to attenuate all the common sources of noise, such as 60-Hz AC power. You can also filter out high-frequency noise above the RF band you are using.

Another way to minimize electrical noise is to use insulated wires and short traces. Even very-small-diameter wires can be purchased with insulation, which often has a clear coating you can strip at the endpoints for soldering. You can rearrange your PCB layout to make your traces short, direct, and well-spaced to reduce noise as well.

You can also create a Faraday cage around your device to protect it from stray noise. A Faraday cage is any type of metal shield (which can be deposited via sputter deposition or made of metal) that creates an enclosure used to block electromagnetic fields. The shield is then grounded and whatever is inside is protected.

Vibration

MEMS sensors and actuators are often used in vibratory environments, such as the sensors on car airbags or on rotating machinery. Even wearable sensors are subjected to vibrations when the person walks or rides in a vehicle. So, the environment in which your device will be used needs to be simulated as part of your testing plan. The key is to make your mounting and packaging robust, so that the sensor is insensitive to its environment. Know what frequencies it will experience and then use a modal analysis to make sure none of the components have a mode of vibration near one of the expected resonant frequencies.

There are vibratory platforms you can use to do accelerated vibratory tests. Place your packaged device on the platform, turn it on at the expected frequency, and see if your sensor still works.

Biointeractions

How will your device interact with the body? If it is a wearable device, will it corrode with sweat? Will it cause an allergic reaction due to the adhesive or materials? If it is implanted, will the body treat it as a foreign body and have an immune reaction? Silicon is a type of glass and is ignored by the body, but if you are making deep brain stimulators for Parkinson's disease, the electrodes "irritate" the brain tissue and are enclosed with scar tissue over time. So, give some consideration to biointeractions, especially chemical and material allergens, and determine what kind of testing you may need to minimize/prevent adverse reactions.

Heat or Cold

Like most materials, silicon expands and contracts with temperature. It is also heated by physical movement and the passage of electricity through it. These effects are usually minor, but should be considered, especially if the operation of the device is dependent on physical movement, like a strain gauge. In those cases, thermal expansion would be read as increased strain, unless some kind of compensating Wheatstone bridge circuit is employed.

Packaging and PCBs also expand and contract. Like vibration, if they move too much, solder and wire connections can break. The human body is a fairly constant, medium temperature, but if your device is a wearable and could be used while someone is snow skiing or could be left in the sun for prolonged exposure, you will need to test for that.

System-Level Performance

In addition to independent tests of components, you will want to do some overall level system performance tests. The following is a reiteration of ideas already presented.

Indicator Signals

Design in ways to give yourself feedback that the system is working. LEDs are tiny and make great indicator signals. They can turn on and off, change to different colors, flash, and even provide Morse code-type messages to you about the status of your device.

Another great indicator is sound or haptic (vibratory) feedback. When light is not appropriate, consider using a "beep" or a buzzing sensation as indicators.

A third indicator, especially useful when prototyping with an Arduino, is written output on your computer. Have the code tell you what step it is at, what input it is looking for, and the value(s) of the output data. The more information you can get on the screen, the quicker you will determine where the system is hanging.

Subsystem Testing

No matter how tempting it may be to hook everything together and test the overall system, you should *test each subsystem separately*. If you are not sure what is causing the problem in a subsystem, be careful to *change only one parameter at a time* and see how that effects the outcome.

You will probably ignore this advice the first time, but eventually you will come to value what you gain from taking a system-level approach, validating each subsystem and each component as you build towards your final goal. The fact is, even if you get extremely lucky and your system works the first time, (1) you will not know why and (2) you will not know which parameters are controlling the outcome.

Intermediate Outputs

Collect and evaluate data at various intermediate steps. Try changing a parameter and see what effect that has on the output. Does it make a difference, or is your system insensitive to that parameter? Knowing what controls your MEMS device design, its circuit, and packaging parameters will lead to the best long-term success. Using this approach, you will know what is important the next time you want to fabricate a batch, and you will not flounder and wonder why it is not working this time when "you did not change anything." Instead, you will be able to adjust the key parameters and get consistently good results.

Animal Testing

Assuming you are creating a Class III medical device, after all of your cleanroom, outsourced, and benchtop testing, you will eventually need to perform some animal testing before you will be allowed to conduct a human trial. For orthopedics, this is often sheep or goats, and there are university and private facilities in a few locations around the United States that will accept studies from industry. These animal research facilities vary in their level of study control and in what animal models they allow. You will need to work with the facility directly or with a university research partner to conduct an animal study, so start researching what will be involved as soon as possible. Key considerations are as follows:

- Animal species (and quantity) you intend to use—the larger the animal, the more expensive the study
- How your device will be implanted, monitored, and extracted
- Whether your device requires continual data monitoring or runs independently
- The duration of your study

It is important to work with a biostatistician to design your study. They will determine how many animals, how much data, and what data needs to be collected in order to show a treatment effect. Publications often refer to "statistical significance," which is based on the probability of being wrong (reported at P less than or equal to 0.05, or 5%, or P less than or equal to 0.01, or 1%). However, other important factors are the power of the study (few animals leads to low power), the correlation coefficient (how well does the data "fit" the regression curve), and your control of other variables that may inadvertently effect the outcome. For this last factor, you often have a "control group" which is identical to the other animals in all ways, except they do not receive the treatment.

Testing Equipment

Equipment for conducting the various tests you need comes from a variety of sources. The following are the main places.

Within NNCI Facility

Microscopes and metrology equipment are readily available within the NNCI facilities. Any measurements you want to take on a MEMS device, including electrical measures, can usually be done in-house.

Available at University

Other types of test, and even some manufacturing at the macroscale, can often be done within the university. For instance, the University of Washington has a full electronic testing lab and a machining lab that can do wire EDM. Universities with on-site hospitals often have EtO sterilization, and they definitely have autoclaves. Still others have veterinary hospitals and animal research facilities staffed by knowledgeable people familiar with sensors and electronic systems who can conduct your study.

Outsourced

Some testing will need to be outsourced, especially things like FEA or wireless signal transmission tests. I also used outsourcing to fabricate my pattern masks and to do the electropolishing. For previous research projects, I outsourced the mounting of microscopic strain gauges onto my specimen, as I was not dexterous enough to bond them and solder on the lead wires. If the NNCI lab you are working with does not know to whom you could outsource a task, often you can find a lead through another unit at the university or even through other industry users at the lab.

Purchased by You

There are times when you will need or want your own test equipment. If you purchase your own, you will know exactly how to operate it, you will have the same calibration every time, and the equipment will always be available to you. Some of the items I have purchased are a power supply, function generator, oscilloscope, soldering iron, and digital multimeter, along with a bunch of handy supplies, like electronic components, wire, dicing tape, CA glue, and solder paste.

Chapter 11
Transporting

Eventually, you will be done with your MEMS device, or at least be ready to transport your work to another location. The best option is always to hand carry, but that is not always feasible. These devices are small and fragile, so below is some advice on how to best prepare for shipping.

Packaging

When you have a wafer to transport, use a dish with a spider, but also enclose some foam to keep the wafer stationary. Tape the dish closed and wrap it in foam or bubble wrap, and then put the dish inside a padded box. This may seem like overkill, but even with these precautions, shipped wafers break easily.

If you have individual MEMS devices, try to adhere them to something for shipment. If not their final package, then pieces of cardboard. Or slip them into small envelopes. There are small gemstone boxes that work well. These are round or square clear plastic containers with a semismooth foam insert. You place your sensor on the foam, and when the lid snaps shut, your device is held snugly against the lid.

Shock or Drop Risk

If you are shipping through the mail, assume your box will get dropped and package accordingly. The most important thing to prevent is movement—within the gemstone box, the gemstone box within the foam, the wafer within the dish, etc. Pack everything with plenty of padding, but make sure the resulting packing tightly holds your inner packaging and that no settling/loosening can occur.

© Springer Nature Switzerland AG 2019

D. Munro, *DIY MEMS*, https://doi.org/10.1007/978-3-030-33073-6_11

Automating Processes

Once your device has graduated from the NNCI facility, you will need to locate a MEMS fabricator. These are higher-volume facilities with automation. They will work with you to redesign your MEMS device, mounting, packaging, or PCB integration as needed to be compatible with higher-volume processes.

They may also need to subcontract certain processes, which means they may put your MEMS device onto a paper strip roll so that it can be picked off by a machine and placed onto a PCB or into your housing (which will most likely be fabricated by a different vendor).

Just remember that your work is done. You are handing off the design to those who can produce the quantities you need for your application.

Chapter 12
Regulatory Approval Pathways

You want to get your medical device approved by the various regulatory bodies so you can begin marketing it for use in patients. After all, that was the whole point of this adventure into MEMS design. All regulatory body requirements are about the same, but they have a few nuanced differences you should be aware of before applying to them.

Also, know that you are venturing into unknown territory. Very few implanted medical devices contain electronics, and even fewer contain MEMS sensors or actuators. So, the regulatory bodies are going to be cautious and demand more evidence for safety and efficacy than with less complex devices.

Your goal should be to use as much literature evidence as you can from other medical devices that can be considered "predicate devices" in some respect and couple that with your own testing and animal trial results to demonstrate safety and efficacy. You do not want to be a de novo, brand-new device, if you can help it, as that approval pathway takes a very long time. The truth is, you are not that unique. With a little research, you can find where your technology has been used in other approved medical devices, and by tracking down each feature, you can build an overall picture of how your medical device is likely to perform based on the other devices' performance.

The following is a discussion on some of the key regulatory bodies. All of this information came from my own reading and is a compilation from various online resources. Thus, it is only accurate in a general sense. For your specific device and application, you will need to consult with a regulatory advisor.

FDA

If you want to market your medical device in the United States, you need to obtain clearance from the FDA. This clearance is obtained from the FDA via a 510(k) submission, also known as premarket notification. Nearly all Class II

© Springer Nature Switzerland AG 2019

D. Munro, *DIY MEMS*, https://doi.org/10.1007/978-3-030-33073-6_12

(medium-risk) devices, as well as a small number of Class I (low-risk) and a few Class III (high-risk) devices, go through the 510(k) process. Most Class III devices, however, obtain market approval via the premarket approval (PMA) process, which is significantly more strenuous than the 510(k).

For the MEMS device that are the subject of this book, you may have questions about which class your device falls into, along with numerous other questions that cannot be answered without some expert assistance.

In a white paper released by Emergo, a medical device consulting firm (https://www.emergobyul.com/services/united-states), they discuss when to contact the FDA about a new device. The FDA is a huge organization, and whether you are a small start-up company developing your first device or a large corporation with lots of experience, it can be intimidating to communicate with the FDA.

There is no right or wrong time to contact the FDA. Part of their mission statement is to "facilitate medical device innovation by advancing regulatory science, providing industry with predictable, consistent, transparent, and efficient regulatory pathways," and the FDA is populated by real people with real expertise who want to help you. On the other hand, although it is always "better to be safe than sorry," you should also know that a record is being made by the FDA of all your communications and advice/recommendations you have received. When you do submit your application, some of those recommendations may come back to haunt you, so wait long enough before initiating communication that you have a clear idea of what your final device is likely to be so that you avoid unnecessarily raising concerns that will become obsolete or irrelevant.

Whenever you have a question to which you cannot find an adequate answer, you can contact the FDA, but do your research first and take advantage of the numerous databases and online information available to you before picking up the phone.

Even when you may think you have all the answers, there are times when it makes sense to obtain FDA's guidance. Emergo suggests some examples of scenarios that should prompt contact with the FDA:

- Early in the development of a novel device or combination product to ensure the correct classification
- Prior to conducting animal studies to support premarket clearance or an IDE (individual device exemption) human clinical study
- When needing clarification about standards, guidance documents, or test methodologies that may apply to your device
- Before proceeding too far into the development of a clinical investigation plan

When you have decided you need assistance from the FDA, start by calling the Division of Industry and Consumer Education (DICE) within the Center for Devices and Radiological Health (CDRH). DICE consists of former FDA investigators, Office of Device Evaluation (ODE) reviewers, and other specialists whose purpose is to provide technical and regulatory assistance to the medical device industry.

You can submit a formal Request for Information, called a "513(g)," asking for the FDA to review and respond in writing regarding the device classification or the requirements applicable to a device. In particular, the FDA can provide the following in response to a "513(g) Request":

- Assessment of the generic type of device
- The particular class of device
- The type of premarket submission required
- Whether a guidance document is available for the device type
- Additional FDA requirements for the device type

For novel devices, you may get classified as a "presubmission" or "Pre-Sub." If you are designated as Pre-Sub, you may request a face-to-face meeting with the FDA at the agency's Washington, DC, offices, a teleconference, or an email response, depending on the type of device and complexity of the issues and questions. This meeting is highly recommended before conducting clinical, animal, or analytical studies or a marketing application for the following situations:

- The device involves a novel technology.
- The device is for a "first of a kind" indications for use.
- The regulatory strategy is not well established.
- You need guidance on specific issues on study protocols.
- You plan to market your device outside the United States (OUS).

If you decide your device is lower risk, and the 510(k) application process is right for you, understand that the 510(k) process is considerably more rigorous than it once was. For instance, several international standards recognized by the FDA have become de facto requirements for 510(k) submissions, including:

- ISO 14971:2007—Risk Management
- IEC 62366—Usability
- IEC 60601-1—Electrical Safety
- IEC 60601-1-2—Electromagnetic Disturbances
- Numerous sterility standards and more

Conducting the required testing for these standards can be extensive, time consuming, and expensive, and trying to do "all" the testing to cover every contingency is a losing battle. It is better to get proper advice and do only the testing that is essential to prove safety and efficacy.

When you submit your 510(k), the FDA typically subdivides the submission to various subreviewers, each of whom is scrutinizing the application for a specific topic or requirement. For example, the FDA now has expert reviewers for biocompatibility, risk management, electrical safety, and others. In the past, simply stating that this work was done or providing a high-level summary was sufficient. Now, the FDA carefully reviews your actual data, which is why you should conduct your testing according to a plan and provide a rationale or protocol as to why that testing and that data is sufficient.

The first step to obtaining FDA 510(k) clearance is to make sure your product qualifies as a medical device. This is not as simple as it may appear—especially in the age of MEMS devices that may communicate wirelessly, are "combination" devices (such as sensors embedded in orthopedic implants), or use innovative technologies.

The definition of a medical device is that it "is intended for use in the diagnosis of disease or other conditions, or in the cure, mitigation, treatment, or prevention of disease, in man or other animals" or that it "is intended to affect the structure or any function of the body of man or other animals, and which does not achieve its primary intended purpose through mechanical action within or on the body of man or other animals and which is not dependent upon being metabolized for the achievement of its primary intended purpose."

Classes of Medical Devices

If your product meets the legal definition of a medical device, then you must determine its classification to determine if it is eligible for 510(k) review. The FDA's classification system divides medical devices into three classes:

- Class I: devices posing the lowest risk to patients or users, usually with no contact with the inside of the body
- Class II: devices that could harm patients or users if used improperly or malfunction, usually with short-term contact with the inside of the body
- Class III: devices that could severely injure or kill patients or users if used improperly or malfunction, often with long-term contact with the inside of the body (<6 months)

For the most part, Class I devices do not require FDA premarket clearance or approval for sale in the United States. But Class I device manufacturers must still register their products with the agency. Most Class II devices, however, must go through the FDA 510(k) process, while nearly all high-risk Class III devices go through the more rigorous premarket approval (PMA) process.

Again according to Emergo, unlike the EU, where market approval is risk-based, market approval in the United States is based on establishing "substantial equivalence" with another cleared device. This means you identify another device already on the market that shares the same intended use and is technologically similar to your device. It is thus critically important to select the appropriate predicate device and build a strong case for substantial equivalence. *Premarket clearance hinges on the success of this argument.*

Look for a device that closely resembles the *technological features* of your product. At a minimum, the basic technology should be the same. For example, a cardiac pacemaker that has a hermetically sealed titanium canister, with a battery and a rhythm management circuit, may be technologically similar to an implanted insulin pump canister that provides pulsed doses. Although the applications are vastly different, the underlying technology may be "substantially equivalent."

When possible, always choose the predicate device that was cleared most recently, since it was cleared under the FDA's newer, more rigorous requirements.

Some of the new requirements you particularly want to pay attention to when designing your test plan are:

Biocompatibility testing—often required for devices or parts of devices to determine the potential toxicity resulting from contact with the user. Biocompatibility requirements are determined by both the type and duration of contact. Read the ISO 10993 standards (recognized by the FDA) for biological evaluation of medical devices, which provide guidance for selecting the necessary tests for your device.

Software or firmware—validation of software and firmware is required for medical devices. Software is almost considered of "moderate" concern, and unless your firmware is very basic, expect it to be considered of "moderate" concern as well.

Clinical data—a study may be required to obtain clinical data in cases where the FDA does not feel the technology of the subject device is close enough to the predicate device.

Usability studies—demonstrates that the end user can read and understand the directions (called IFUs or instructions for use) and use the device correctly. Usability studies require protocols and well-written reports, but are not considered clinical research and you are usually better off subcontracting the usability study to a third party who specializes in these studies.

510(k)

A 510(k) is commonly broken up into 20 individual sections, each addressing a specific 510(k) requirement, plus all of the relevant protocols, test reports, and other such documentation provided as attachments, so a typical 510(k)s is well over 100 pages.

EU

Big changes are underway for the European market and how it regulates medical devices with the recent unveiling of the Medical Device Regulation MDR 2017/745. This regulation replaces the long-standing Medical Devices Directive (MDD) 93/42/EEC, specifically MedDev 2.7/1 for medical devices, which was created in 1992 and most recently updated to Revision 4 in July 2016. The MDR is a significant deviation from the MDD and MedDev 2.7/1, and it is important to implement necessary changes to how one conducts literature searches, postmarket surveillance (PMS), and postmarket clinical follow-up (PMCF) in the 2017–2018 timeframe to be prepared for MDR by 2020. The following is a set of guidelines for making this transition smoothly.

European Single Market and Medical Devices

The European Single Market has 28 member states of the European Union (including the UK), the European Economic Area (Liechtenstein, Norway, and Iceland), and Switzerland (through bilateral treaties). The purpose of the European Single Market is to allow free movement of goods from one member state's market to another, assuming the following three criteria have been met:

- Essential requirements for the products involved have been defined.
- Methods that describe how product compliance with the requirements have been established.
- Mechanisms to supervise and control the actions of those involved in manufacturing and distributing the product have been created.

To address these criteria, the Medical Devices Directive (MDD) 93/42/EEC was created in 1992 and was most recently revised in 2009 (Revision 3) and 2016 (Revision 4). However, the MDD has some inherent weaknesses that make interpretation of the directives inconsistent across the member states, which allowed some defective products to be released into the market. In addition, the roles of the competent authorities and numerous Notified Bodies were unclear, and the enforcement of the directives across the member states was haphazard.

While attempting to address these weaknesses, it became clear that the MDD could not be revised sufficiently to correct the directive, and so a new document was drafted that combined the MDD, the *Active Implantable Medical Devices Directive*, and the In Vitro *Diagnostic Regulations* into a single document that also included software and accessories in its definition of medical devices. The MDR was published in the *Official Journal of the European Union* on May 2017, and full implementation is expected to occur in 2020.

The key difference between the MDD and the MDR is the requirement for clinical evaluations. Equivalence will become an increasingly difficult means by which to demonstrate compliance, and Class III devices will be expected to do clinical evaluations. Strict rules for clinical investigations and alignment to Clinical Trial Regulations are introduced in Chapter VI, Articles 62 to 82 for the MDR.

MedDev 2.7/1 Revision 4 as a Pathway to MDR

MedDev 2.7/1 Revision 4 is a good first step for the transition to MDR, as it implements many of the changes in regards to establishing equivalence and collecting clinical data. It is also important to note that all devices approved under Revision 4 of the MDD that are lawfully already placed on the European market will be grandfathered in as in compliance with the MDR and will not have to perform clinical investigations to continue being sold on the European market. However, these products will have to update their clinical evaluation reports (CERs) to adhere to the new requirements of the MDR in all other ways.

So, let us first explore Revision 4 and how it differs from Revision 3 of MedDev 2.7/1. The first thing one notes is that it is significantly longer than Revision 3, being 65 pages instead of 46. Most of the increase is due to more detailed and expansive requirements for clinical data. It is the goal of Revision 4 to clarify what is meant by evaluation of clinical data and to provide more detailed examples of where one should seek clinical data. It also provides templates for manufacturers to follow and provides guidelines for Notified Bodies.

Revision 4 places a lot of emphasis on conducting clinical evaluation throughout the life cycle of a medical device, including its design stage. The CER should be updated annually per Revision 4, which means all of the supporting reports that drive change must be updated annually as well. Specifically, the literature search results and the postmarket surveillance (PMS) reports will need to be conducted annually. There is an expectation in Revision 4 that postmarket clinical follow-ups (PMCF) will also be conducted, but it is not stated that these follow-ups are required or that they must be done on an annual basis.

Revision 4 provides guidance on where and how to search for literature. It is necessary for the manufacturer to have a governing literature search protocol that is developed and approved by the manufacturer prior to conducting the literature search. This document must be cited in the CER. In addition, the full results of the literature searches must be kept in a file that is available for the Notified Body to review. It must show which articles were selected and which were rejected based on the criteria established in the protocol. There is further guidance on how to analyze each article found and what types of articles should be rejected.

It is still possible to use data from an equivalent device; however, the definition of "equivalence" has been narrowed and refined, especially as compared to the FDA's definition of "substantial equivalence." For Revision 4, the equivalent device must be the same geometry, material, and function as the manufacturer's device. If only two of the criteria are met, say function and geometry, but the material is a different chemical composition, then it will be difficult to use as equivalent. The evaluator would have to justify why the device could still be used. Thus, multiple literature searches will have to be performed to find equivalence for each feature of a device. It is further necessary to find pictures of the equivalent devices for each feature claimed as equivalent.

Revision 4 requires transparency from manufacturers on their methods used to gather data about their device. This means all data collected must be logged in raw form, and methods used to analyze the raw data must be documented in a report. Thus, the CER will reference multiple supporting documents and will contain summarized tables and/or reports of the findings of the supporting documents.

Revision 4 makes compliance with the essential requirements for safety, performance, and risk-benefit analysis a priority. The new template provided for Revision 4 provides a separate subsection for each of these topics so that the manufacturer specially addresses each one.

There are also more detailed instructions on how to establish and document the state-of-the-art and available treatment options. The CER should include an introduction and background to the field in which the device is used, and therefore, a literature search specifically for finding this data will be necessary. The manufacturer will need to address the risk and benefits of the various treatment options available.

Revision 4 requires that the Notified Body challenge the manufacturer's claims of equivalence with other devices. The Notified Body will thus need access to all of the log data so they can review the assumptions and choices made by the evaluator. This is a significant deviation from Revision 3, but is a foreshadowing of things to come in MDR 2017/745, where the Notified Body's role changes from collaborator with the manufacturer to a policing force.

In Revision 4, the links between clinical evaluation, PMS, and PMCF reports are now the driving forces for the CER updates. It will therefore be critical for a PMS system to be established and reported upon annually at least 2 months prior to a CER update. The same is true for a PMCF system, although this could probably be most easily accomplished by gathering survey data from representatives present during the surgical procedure and recording it in a database.

There is now a specific requirement in Revision 4 for demonstrating the scientific validity of all data found, including statistical considerations. Percentages are no longer considered sufficient for analysis, so the data should be evaluated with statistics, such as mean, standard deviation, and P-value, when making claims about safety, performance, or risk-benefit of a medical device.

Finally, there are now specific requirements for the expertise and experience of CER authors and evaluators. Postgraduate education and 5 years of experience in the field or 10 years of experience in the field and justification of one's expertise are now required. In addition, each evaluator must make a Declaration of Interest, which discloses their affiliation with the manufacturer.

Medical Device Regulation Preparations

The most important take away for manufacturers to prepare for the MDR is that they need to define protocols for and start gathering clinical and market data now for all their medical devices, software, and accessories. These protocols are used to generate the literature search, clinical trial, PMS, and PMCF reports. Below is an outline of the key protocols which need to be written. Reports will then be generated annually based on these protocols, and the collective results will be used to drive the CER.

- *Literature search protocol* with weighting and selection criteria for articles, full database of raw results, and a literature search results report. These three items can be compiled into one report with all the results attached, but it would make more sense to keep them separate, as the raw data and results report need to be updated annually.
- *Postmarket surveillance (PMS) report* that routinely monitors various other types of data, both internally and externally generated, such as incident reports, MAUDE, national joint registries, and feedback from users, for issues

or problems with the Instructions for Use (IFUs), accessories, devices, or marketing materials associated with the product.

• *Postmarket clinical follow-up (PMCF) report* that monitors performance and safety of the device in the surgical arena, such as incident reports, MAUDE, national joint registries, feedback from users, and surveys completed by representatives in the operating room, for issues or problems with the device's clinical performance.

The purpose of the CER then becomes a decision-making document. Based on what is found in the literature, PMS, and PMCF reports, what changes or improvements need to be made to the device? What plans will be implemented to make the needed changes if any are identified during the CER? If this is a new medical device, what was learned from the clinical trial about the device's performance?

All accessories used with or for a medical device, even if they have no intended medical purpose, are now considered medical devices. The definition of an accessory is expanded to "assist" a device to be used (before, the definition said "enable" a device to be used). Another significant increase of scope is that devices for cleaning, disinfecting, or sterilizing a device will also be classified as medical devices. Thus, it is my interpretation that all surgical instruments and implant trials used during a surgical procedure would be classified as medical devices. The tray and the sterilization techniques for devices and instruments would be regulated by their manufacturer or operator, but the packaging and packaging systems would need to be monitored with a database maintained by the manufacturer. The Notified Body will be involved with regulating sterility of all packaging and packaging systems, too. If all surgical instruments and implant trials are going to be part of the CER, this will significantly increase the scope of the document.

How this regulation will impact combination devices, such as MEMS used in traditional medical devices, is yet to be determined.

EUDAMED and UDI

Additionally, each device will require a unique device identification (UDI), and there will be a uniform, international database where the manufacturer will have to maintain their UDI information and cross-reference it to its Declaration of Conformity. This database, called EUDAMED, will be accessible to various users, some of it even to the individual patient. EUDAMED must also be used to document PMCF studies. There remain open questions about how this will impact privacy laws and security concerns. EUDAMED will require member states to issue unique Single Registration Numbers to each EUDAMED user. This is expected to be a complex and demanding effort for which no resources have been identified yet.

Manufacturers will be required to supply their competent authority with all information necessary to demonstrate conformity with the MDR, as well as share that information with patients or their representatives claiming compensation. An additional part of the requirement is the UDI and a new label of either "MD" for medical device and "IVD" for in vitro device.

Equivalence and Establishing Predicate Devices

For the device being evaluated, it is important to compile a detailed list of all components and features for the system. This needs to include the sizes, materials, fabrication or surface treatment methods, unique or special features, indications for use, and any other differentiators that will impact the literature search.

This list will help identify equivalent devices and their manufacturers. The list of equivalent devices and device features need to be compiled into a table that contains all of the requisite information:

- Exact names, models, sizes, materials, coatings, treatments, features, accessories, etc.
- Name of the manufacturer
- Relationship to the device under evaluation (predecessor/successor, others)
- Regulatory status
- If the device is not CE-marked, justification for the use of the data
- Comparison of clinical, biological, and technical characteristics
- Description of relevant clinical, biological, and technical characteristics that affect clinical properties of the device
- Differences between the intended purpose of the device under evaluation and the equivalent device (indications, contraindications, precautions, target patient groups, target users, mode of application, duration of use/number of reapplications, others) and type of device-body interactions
- Choice, justification, and validity of parameters and models for nonclinical determination of characteristics

This in turn guides the literature search, as some article and data provide more justification for equivalence than others, such as

- Identification of preclinical studies carried out and literature citations
- Clear studies and literature for the equivalent device or feature (methods, results, conclusions of the authors)
- Methodological quality of the study or document
- Scientific validity of the information

Results must then be tabulated in a concise and easy to follow manner:

- Comparative tabulations for the device under evaluation versus the equivalent device showing parameters relevant to the evaluation of the three characteristics (material, geometry, use)
- Comparative drawings or pictures of the device and the equivalent device showing the elements in contact with the body
- Identification of differences and an evaluation if differences are expected or not to influence the clinical performance and clinical safety of the device, with reasons for assumptions made

The tables should correlate with brief summaries for each equivalent device that explain:

- Whether the comparison carried out covers all products, models, sizes, settings, accessories, and the *entire* intended purpose of the device under evaluation.
- Or if only *certain* products, models, sizes, settings, accessories, or selected aspects of the intended purpose are considered equivalent.
- Conclusions and justification for whether equivalence is demonstrated or not; if it is demonstrated, confirmation that the differences are not expected to affect the clinical performance and clinical safety of the device under evaluation; if it is not demonstrated, include a description of any limitations and gaps.

Safety

The goal of regulatory review is safety first. It is the Hippocratic Oath that all doctors and caregivers pledge to "Do No Harm." Keep this in mind when performing your testing and evaluations, and you will have confidence that your device will not harm anyone. You will read the literature, compare your device to other equivalent devices, and perform preclinical and clinical testing. Then, you will determine whether (1) the risks identified in your documentation and the literature have been adequately addressed, (2) all the hazards and other clinically relevant information have been identified appropriately, and (3) the safety characteristics and intended purpose of the device (including training of the end user or other precautions) are described in the IFU.

For safety, the primary outcome measure in orthopedics is revision, with revision meaning the replacement of a prosthetic component, but your medical device may have a different measure. Regardless, to assess performance outcomes, you should:

- Identify the appropriate national registries for your device and review them
- Decide if your device and identified predicates are performing as excepted
- Document the specific reasons for revision (or your outcome measure)
- Look for indicators that could predict late failure may occur, such as loosening of a connection or debonding on a package from the implant

Since MEMS devices are relatively new, there may not be much data available, which is why testing and measurement throughout your design process will be critical. And it will be equally important to find technologically similar medical electronics to provide context for reviewers.

Efficacy

Your device may be completely safe, but if it provides no real benefit to the patient, the risks associated with surgery or implantation do not warrant its use. Like any new technology, everyone jumps on board to sell a product using it. I went to the Consumer Electronics Show (CES) with 50,000 other people in Las Vegas, and rather than feel impressed, I was stunned by the number of completely useless gadgets that were being produced. There were a lot of cool "solutions looking for a problem." The same phenomenon is occurring in medical devices using MEMS technology, unfortunately.

When deciding whether or not to incorporate MEMS into your product, think about how it will benefit the end user. Will it provide additional information that cannot be determined any other way? Will it alert someone to a change in conditions that could improve outcomes? Will it provide data in a less invasive or less harmful way? Will it be a paradigm shift in how we diagnose or treat a condition? If yes, then it is worth the time to develop a MEMS solution.

Chapter 13
Locations of US Facilities

Description of Locations

The National Nanotechnology Coordinated Infrastructure (NNCI) consists of 16 primary sites distributed across the United States. Of these, two are specialized research facilities and the rest are open use facilities that welcome industry users.

On the west coast, the facilities are the University of Washington (with partner Oregon State University), Stanford University, and the University of California at San Diego.

In the western region, the facilities are Montana State University (with partner Carlton College), Arizona State University (with partners Maricopa County Community College and Science Foundation Arizona), University of Nebraska at Lincoln, and the University of Texas at Austin.

In the Midwestern region, the facility is the University of Minnesota at Twin Cities with partner North Dakota State University. There is also a specialized facility for soft and hybrid nanomaterials at Northwestern University with partner University of Chicago. Not part of NNCI, but also at Northwestern University, is the Nanotechnology Corporate Partners (NCP) Program, which collaborates with industry.

In the southern region, the facilities are the University of Louisville (with partner University of Kentucky) and Georgia Institute of Technology (with partners North Carolina A&T State University and University of North Carolina at Greensboro).

On the east coast, the facilities are Harvard University, Cornell University, the University of Pennsylvania, and North Carolina State University (with partners Duke University and University of North Carolina at Chapel Hill). There is also a specialized facility for Earth and environmental science at Virginia Tech (with partner Pacific Northwest National Laboratory [PNNL]).

© Springer Nature Switzerland AG 2019
D. Munro, *DIY MEMS*, https://doi.org/10.1007/978-3-030-33073-6_13

Tabulated List of Locations

Each site submitted an application with a detailed abstract. This information can be found at http://www.nnin.org/news-events/news/nnci-award and is summarized here by region with all of the key contact information. It is further tabulated in Table 13.1.

Table 13.1 Summary of NNCI facilities

NNCI facility name	Primary site	Location	NNCI site name	Facility website
West coast region				
Northwest Nanotechnology Infrastructure	University of Washington	Seattle, Washington	NNI	www.wnf. washington.edu
Nano@Stanford	Stanford University	Palo Alto, California	NanoStanford	nanolabs.staaford. edu
San Diego Nanotechnology Infrastructure	University of California at San Diego	La Jolla, California	SDNI	sdni.ucsd.edu
Western region				
Montana Nanotechnology Facility	Montana State University	Bozeman, Montana	MONT	www.nano. montana.edu
Nanotechnology Collaborative Infrastructure Southwest	Arizona State University	Tempe, Arizona	NCI-SW	ncisouthwest.org
Nebraska Nanoscale Facility	University of Nebraska at Lincoln	Lincoln, Nebraska	NNF	nanoscale.unl.edu
Texas Nanofabrication Facility	University of Texas at Austin	Austin, Texas	TNF	www.mrc.utexas. edu
Midwestern region				
Midwest Nano Infrastructure Corridor	University of Minnesota at Twin Cities	Minneapolis, Minnesota	MiNIC	minic.umn.edu
Soft and Hybrid Nanotechnology Experimental Resource	Northwestern University	Evanston, Illinois	SHyNE	www.shyne. northwestern.edu
Nanotechnology Corporate Partners (NCP) Program	Northwestern University	Evanston, Illinois	*Allows corporate partners	www.iinano.org/ nanotech-nology-corporate-partners-program

(continued)

Table 13.1 (continued)

NNCI facility name	Primary site	Location	NNCI site name	Facility website
Southern Region				
Kentucky Multi-scale Manufacturing and Nano Integration Node	University of Louisville	Louisville, Kentucky	KY-MMNIN	www. kymultiscale.net
Southeastern Nanotechnology Infrastructure Corridor	Georgia Institute of Technology	Atlanta, Georgia	SENIC	senic.gatech.edu
Eastern Region				
Cornell Nanoscale Science and Technology Facility	Cornell University	Ithaca, New York	CNF	www.cnf.cornell. edu
Virginia Tech National Center for Earth and Environmental Nanotechnology	Virginia Polytechnic Institute and State University	Blacksburg, Virginia	NanoEarth	www.nanoearth. ictas.vt.edu
North Carolina Research Triangle Nanotechnology Network	North Carolina State University	Raleigh, North Carolina	RTNN	www.rtnn.ncsu. edu
East coast region				
Center for Nanoscale Systems	Harvard University	Cambridge, Massachusetts	CNS	cns1.rc.fas. harvard.edu
Mid-Atlantic Nanotechnology Hub	University of Pennsylvania	Philadelphia, Pennsylvania	MANTH	www.nano.upenn. edu

West Coast Region

Northwest Nanotechnology Infrastructure (NWNI)

Investigator(s):	Karl Bohringer karl@ee.washington.edu (principal investigator) Lara Gamble (co-principal investigator) Daniel Ratner (co-principal investigator) W. James Pfaendtner (co-principal investigator) Daniel Schwartz (co-principal investigator)
Sponsor:	University of Washington 4333 Brooklyn Ave NE Seattle, WA 98195-0001 (206)543-4043

Abstract

The Northwest Nanotechnology Infrastructure (NWNI) as an NNCI site serves as the prime resource for nanotechnology researchers and engineers for a large geographical area from the Pacific Coast to Montana and from southern Oregon to the Canadian border and beyond. NWNI offers world-class facilities at the University of Washington (UW) in Seattle and Oregon State University (OSU) in Corvallis, complemented with unique capabilities at Pacific Northwest National Laboratory, a Department of Energy site, and the University of British Columbia in Vancouver, Canada. Anchored at the UW, this site provides critical workhorse tools, unique instruments, and key educational support to a large and distributed user base with particular attention to the clean energy and biotechnology fields.

The mission of NWNI consists of four core services that can be described by four Ms: make, measure, model, and mentor. The first three Ms form the physical foundation and the fourth serves to coordinate educational efforts with broad impact beyond the scientific community. The physical infrastructure consists of the Washington Nanofabrication Facility (WNF, Seattle) and the Microproducts Breakthrough Institute (MBI, Corvallis) for making, the Molecular Analysis Facility (MAF, Seattle) and the Materials Synthesis and Characterization Facility (MaSC, Oregon) for measuring and distributed computational resources for modeling in design and analysis. Mentoring is essential to NWNI. The site's integration with the region's vibrant biotech and start-up community implies immense diversity in users.

NWNI offers flexible access to its facilities, from comprehensive training of local users to operator-assisted tool access to remote execution of assignments. Whether novice or seasoned engineer or scientist, whether undergraduate, graduate, postdoc, or community college student or teacher, all users are offered support for their entire nanotechnology project from initial design to final analysis.

The NWNI serves as a broad-based nanotechnology resource, though three principal research focus areas are highlighted in which the site will provide leadership:

  (i) Integrated photonics, which aims at enabling large-scale photonic networks, which are expected to overcome current limits in speed and bandwidth of electronic circuits. Beyond information processing, the miniaturization and integration of photonics in medical devices is facilitating the development of new, minimally invasive health diagnostics.

 (ii) Advanced energy materials and devices, which aims at providing the scientific and engineering basis for clean energy solutions, including the creation of better batteries or scalable and environmentally benign materials for solar power.

(iii) Bio-nano interfaces and systems, which provides the infrastructure and expertise for inventing and demonstrating new devices for biomedical applications, enabling advances in protein modeling, drug delivery, sensors, bio-scaffolds, and bioelectronics. NWNI features capabilities in materials and devices including quantum dots, superabsorbers for solar cells, and oxide-base transistors for flexible electronics for sensors and displays, resulting in comprehensive infrastructure and expertise in nanotechnology that is considered unique within the field.

The site provides an array of educational activities geared toward a broad audience and designed to have a multiplier effect. Three signature residence programs are offered:

(i) Educators in residence gain hands-on laboratory skills for use in teaching their K-12 classes.
(ii) Entrepreneurs in residence, in coordination with UW Technology Transfer (CoMotion), work with nanotechnology inventors to explore start-up opportunities.
(iii) OSU Advantage connects businesses with faculty expertise, student talent, and world-class NWNI facilities to assist in bringing ideas to market.

In a collaboration with the University of British Columbia, online edX courses are supported that allow students across the country and around the world to build and test their own nanoscale photonic devices on multiproject wafers built with electron beam lithography at UW. Worldwide, UW participates in a network of 15 institutions in America, Asia, and Europe that offer joint summer schools on nanotechnology for future global engineers.

Stanford University—SNSF, SNF, MAF, EMF

Investigator(s):	Bruce Clemens clemens@soe.stanford.edu (principal investigator)
	Kathryn Moler (former principal investigator)
	Beth Pruitt (co-principal investigator)
	Curtis Frank (co-principal investigator)
	Katharine Maher (co-principal investigator)
Sponsor:	Stanford University
	3160 Porter Drive
	Palo Alto, CA 94304-1212 (650)723-2300

Abstract
The Stanford site of the National Nanotechnology Coordinated Infrastructure (NNCI) at Stanford University will provide open, cost-effective access to state-of-the-art nanofabrication and nanocharacterization facilities for scientists and engineers from academia, small and large companies, and government laboratories. Stanford will open the Stanford Nano Shared Facilities (SNSF), the Stanford Nanofabrication Facility (SNF), the Mineral Analysis Facility (MAF), and the Environmental Measurement Facility (EMF) more fully to external users.

Open access to these facilities will not only promote the progress of science but also accelerate the commercialization of nanotechnologies that can solve a broad array of societal problems related to energy, communication, water resources, agriculture, computing, clinical medicine, and environmental remediation. Stanford will create and assemble a comprehensive online library of just-in-time educational materials that will enable users of shared nanofacilities at Stanford and elsewhere to acquire foundational knowledge independently and expeditiously before they receive personalized training from an expert staff member. Stanford staff members will also collaborate with two minority-serving institutions (California State University Los Angeles and California State University East Bay) to provide coursework, hands-on training, and nanofacility access to their students.

The Stanford site's shared nanofacilities will offer a comprehensive array of advanced nanofabrication and nanocharacterization tools, including resources that are not routinely available, such as an MOCVD laboratory that can deposit films of GaAs or GaN, a JEOL e-beam lithography tool that can inscribe 8-nm features on 200-mm wafers, a NanoSIMS, and a unique scanning SQUID microscope that detects magnetic fields with greater sensitivity than any other instrument.

The facilities occupy ~30,000 ft^2 of space, including 16,000 ft^2 of cleanrooms, 6000 ft^2 of which meet stringent specifications on the control of vibration, acoustics, light, cleanliness, and electromagnetic interference. The staff members who will support external users have acquired specialized expertise in fabricating photonic crystals, lasers, photodetectors, optical MEMS, inertial sensors, optical biosensors, electronic biosensors, cantilever probes, nano-FETs, new memories, batteries, and photovoltaics.

Stanford will endeavor to increase the number of users from nontraditional fields of nanoscience (e.g., life science, medicine, and Earth and environmental science) by creating a targeted formal curriculum, fabricating experimental nanostructures as a service, providing seed grants, and leading seminars and webinars.

San Diego Nanotechnology Infrastructure (SDNI)

Investigator(s):	Yu-Hwa Lo ylo@ece.ucsd.edu (principal investigator)
	Shaochen Chen (co-principal investigator)
	Eric Fullerton (co-principal investigator)
	Yeshaiahu Fainman (co-principal investigator)
Sponsor:	University of California-San Diego
	Office of Contract and Grant Admin
	La Jolla, CA 92093-0621 (858)534-4896

Abstract

The San Diego Nanotechnology Infrastructure (SDNI) site of the NNCI at the University of California at San Diego offers access to a broad spectrum of nanofabrication and characterization instrumentation and expertise that enable and accelerate cutting-edge scientific research, proof-of-concept demonstration, device and system prototyping, product development, and technology translation. Nanotechnology is the cornerstone of many industry sectors and a rich source for scientific discoveries and innovations. Using nanotechnologies, scientists are likely to find solutions for the most important challenges in health, communications, energy, and environment.

Nanotechnology is multidisciplinary by nature and requires highly sophisticated tools and deep expertise, often unavailable or unaffordable by individual research labs and businesses. The SDNI site will offer state-of-the-art knowhow, tools, and services of nanotechnologies to all interested users across the nation in a user-friendly, timely, and cost-effective manner. The site will also become a nanotechnology provider to create and develop new nanotechnologies and bring them to its users.

The goals of the site are to serve a large number of academic, industrial, and government users, to transfer enabling nanotechnologies from research laboratories to the general user community, to educate and train future generations of scientists and engineers in nanotechnology, and to bring nanoscaled research experience to college students and K-12 students, especially underrepresented minority students, to prepare them for STEM careers.

The SDNI site will build upon the existing Nano3 user facility and leverage additional specialized resources and expertise at the University of California at San Diego. The SDNI site is committed to broadening and further diversifying its already substantial user base. The proposed strategic goals include:

(i) Providing infrastructure that enables transformative research and education through open, affordable access to the nanofabrication and nanocharacterization tools and an expert staff capable of working with users to adapt and develop new capabilities, with emphasis in the areas of nanobiomedicine, nanophotonics, and nanomagnetism

(ii) Accelerating the translation of discoveries and new nanotechnologies to the marketplace

(iii) Coordinating with other NNCI sites to provide uninterrupted service and creative solutions to meet evolving user needs

Significant growth is anticipated in the number and variety of local and regional users in the academic, government, and industrial sectors. Discoveries made by users of the SDNI site have the potential to create transformative change in fields as diverse as medicine, information technology, transportation, homeland security, and environmental science, leading to improved healthcare, faster communications, safer transit, and cleaner water and air.

To develop a more diverse and productive scientific workforce, the SDNI site will expand undergraduate and graduate training programs including REU opportunities to train 900 students over 5 years. Through an RET program and other activities, the site will work to increase the number of students from underrepresented minority groups who pursue studies and, ultimately, careers in STEM disciplines.

Western Region

The Montana Nanotechnology Facility (MONT)

Investigator(s):	David Dickensheets davidd@ee.montana.edu (principal investigator)
	Recep Avci (co-principal investigator)
	David Mogk (co-principal investigator)
	Philip Stewart (co-principal investigator)
Sponsor:	Montana State University
	309 Montana Hall
	Bozeman, MT 59717-2470 (406)994-2381

Abstract

Nanotechnology, which gives us the ability to manipulate and interrogate physical systems on a length scale of nanometers to microns, has become pervasive in many fields of scientific inquiry and engineering. Access to basic nanotechnology tools has therefore become increasingly important, not only for so-called nanotechnologists but for scientists and engineers from many academic disciplines and from industry.

The Montana Nanotechnology Facility (MONT), an NNCI site at Montana State University, promotes discovery, education, and outreach related to nanotechnology by providing access to shared-use instruments, expert training on their safe and effective use, and broad-based education about nanoscale science and technology for learners at all levels who come from diverse communities.

The MONT site serves both regional users in the northern Rocky Mountains and Great Plains and users from across the United States who need the specific expertise and equipment found at Montana State University. Those users are pursuing diverse objectives related to advances in healthcare diagnostics and surgical solutions, sources of clean energy, remediation strategies for contaminated soils, and technologies related to optical telecommunications, imaging systems, and advanced computing.

By enhancing our service to external users and building on its unique fabrication and characterization strengths, MONT will help to meet a national need for access to nanotechnology, for training of the workforce that will develop the nanotechnology of the future, and for education and outreach that engages and informs students and teachers from kindergarten to graduate school, industrial users, and the general public.

MONT helps meet the growing need faced by regional and national researchers for access to nanofabrication tools and processes at the interdisciplinary frontiers, with local expertise related to microelectromechanical systems (MEMS) and micro-opto-electromechanical systems (MOEMS); microfluidics; nanostructured materials with unique optical, mechanical, or thermal properties; ceramic materials; bio-inspired and bio-derived nanostructures; and bacteria or bacterial biofilms incorporated into micro- or nanoengineered substrates. The goals of the MONT site are as follows:

(i) To increase the number of external users served

(ii) To increase the collective research output of MONT users

(iii) To enhance the MONT site's capabilities in the areas of its research strengths through heavily leveraged capital investment

(iv) To create best-in-class educational opportunities for facility users, STEM educators, and the general public

These goals are accomplished through specific initiatives that will add laboratory personnel to enhance training, assistance, and advocacy for external users, establish a user grant program for external users to help address costs of facility use as well as local housing, invest in new tools and capabilities, and expand both on-site and web-based instructional and outreach activities related to nanofabrication, nano-characterization, and the ethics and societal impacts of nanotechnology.

The project specifically improves access to nanotechnology infrastructure in the northern Rockies/Great Plains region, and it promotes discovery, education, and outreach in emerging fields where nanotechnology is impacting the life sciences, healthcare, energy, the environment, and a number of important technology sectors.

Nanotechnology Collaborative Infrastructure Southwest (NCI-SW)

Investigator(s):	Trevor Thornton t.thornton@asu.edu (principal investigator)
	Stuart Bowden (co-principal investigator)
	Jameson Wetmore (co-principal investigator)
	Jenefer Husman (former co-principal investigator)
Sponsor:	Arizona State University
	ORSPA
	TEMPE, AZ 85281-6011 (480)965-5479

Abstract

Arizona State University (ASU) will establish the Nanotechnology Collaborative Infrastructure Southwest (NCI-SW) as an NNCI site. The NCI-SW will support the advanced toolset, faculty expertise, and knowledgeable staff required by academic and industrial users performing research at the frontiers of nanoscience and engineering.

Its training programs will focus on workforce development and entrepreneurial initiatives for twenty-first century manufacturing industries. A partnership between ASU, Maricopa County Community College District (MCCCD), and Science Foundation Arizona (SFAz) will allow two-year colleges in metropolitan Phoenix and rural Arizona to deliver a STEM-based nanotechnology curriculum designed to meet the economic development needs of their communities.

Particular emphasis will be placed on programs in rural Arizona that support Hispanic and Native American students. Students in these programs will have access to advanced laboratory facilities either directly on the ASU campus or via remote access. Faculty and students from local high schools and community colleges will collaborate with ASU faculty on summer research programs at the frontiers of nanotechnology and develop lesson plans that convey the excitement of the latest discoveries back to their classrooms. Public outreach events at science fairs and at the Arizona Science Center will allow the wider community access to the latest breakthroughs in nanotechnology at ASU and from around the world.

The goals of the NCI-SW are to build a southwest regional infrastructure for nanotechnology discovery and innovation, to address societal needs through education and entrepreneurship, and to serve as a model site of the NNCI. The NCI-SW site will encompass six collaborative research facilities: the ASU NanoFab, the LeRoy Eyring Center for Solid State Science, the Flexible Electronics and Display Center (FEDC), the Peptide Array Core Facility, the Solar Power Laboratory (SPL), and the user facility for the social and ethical implications of nanotechnology.

The NCI-SW site will open the FEDC and SPL to the broader research community for the first time. The site will provide particular intellectual and infrastructural strengths in the life sciences, flexible electronics, renewable energy, and the societal impact of nanotechnology. ASU will collaborate with Maricopa County Community College District (MCCCD) and Science Foundation Arizona (SFAz) to develop STEM materials with a nanotechnology focus for AS and AAS students in communities throughout metropolitan Phoenix and rural Arizona. NCI-SW will provide entrepreneurship training for users who wish to commercialize nanotechnology in order to benefit society.

To facilitate the commercialization of research breakthroughs, the NCI-SW will support prototyping facilities and low-volume manufacturing pilot lines for solar cells, flexible electronics, and biomolecular arrays. The Science Outside the Lab summer program at the ASU Washington, DC, campus will allow users across the NNCI to explore the policy issue associated with nanotechnology.

A web portal hosted and maintained by MCCCD will provide seamless access to all the resources of the NCI-SW.

Nebraska Nanoscale Facility (NNF)

Investigator(s):	David Sellmyer dsellmyer1@unl.edu (principal investigator)
	Rebecca Lai (co-principal investigator)
	Christian Binek (co-principal investigator)
	Sy-Hwang Liou (co-principal investigator)
	Jeffrey Shield (co-principal investigator)
Sponsor:	University of Nebraska-Lincoln
	151 Prem S. Paul Research Center
	Lincoln, NE 68503-1435 (402)472-3171

Abstract

The Nebraska Nanoscale Facility (NNF) at the University of Nebraska will provide a regional center of excellence for instrumentation and service in nanoscience and nanotechnology to the NNCI. It will contribute to the US research and educational infrastructure for transformative advances in the fabrication, understanding, and utilization of novel nanostructures, materials, and devices.

These structures and devices play an increasingly critical role in contemporary technologies including ultraminiaturization in information processing, digital communications, energy processing, sensors for threat detection, and biomedicine. Special attention will be given to serving the nanotechnology needs of educational institutions and industry in the western region of the US Midwest.

NNF will significantly enhance economic development through industrial collaborations, spin-offs, materials analyses, and tech transfer to companies. National impact will result from interactions and collaborations with the newly developing Nebraska Innovation Campus and the National Security Research Institute at the University of Nebraska. A strong education-outreach program at NNF is focused on increasing diversity through summer research experiences for students and professor-student pairs, after-school middle-school programs, community college programs, minicourses, and others. In addition, education and outreach efforts will be pursued with Native Americans and tribal colleges in Nebraska associated with the Winnebago, Santee Dakota, and Omaha tribes.

NNF will build upon the established Central Facilities of the Nebraska Center for Materials and Nanoscience to strongly galvanize research and education in nanotechnology in Nebraska and the region. The Central and Shared Laboratory Facilities include: nanofabrication cleanroom, nanomaterial and thin-film preparation, nano-engineered materials and structures, electron microscopy, X-ray structural characterization, scanning probe and material characterization, low-dimensional nanostructure synthesis, and laser nanofabrication and characterization. Most of these facilities are housed in the 32,000 sq. ft. Voelte-Keegan Nanoscience Research Center that was completed in 2012 and funded by major grants from the National Institute for Standards and Technology and the University of Nebraska Foundation.

The research in NNF is bolstered by strong research groups in nanoscale electronics, magnetism, and materials and structures for energy. NNF in turn will reinforce several centers and focused research programs including the Nebraska NSF-MRSEC: Polarization and Spin Phenomena in Nanoferroic Structures, DOE-EERE Consortium on Magnetic Materials, SRC-NIST Center for Ferroic Devices, NSF-Center for Nanohybrid Materials, and others. These programs have many national and international collaborators that will add vitality to and provide a broad base of users for the NNF. Hundreds of graduate and undergraduate students, post-doctoral research associates, and visiting scientists and engineers from companies will benefit each year from the state-of-the-art facilities in NNF.

Texas Nanofabrication Facility (TNF)

Investigator(s):	Sanjay Banerjee banerjee@ece.utexas.edu (principal investigator)
	Lee Kahlor (co-principal investigator)
	Arumugam Manthiram (co-principal investigator)
	S. Sreenivasan (co-principal investigator)
	Keith Stevenson (former co-principal investigator)
Sponsor:	University of Texas at Austin
	3925 W Braker Lane, Ste 3.11072
	Austin, TX 78759-5316 (512)471-6424

Abstract

Nanotechnology deals with man-made objects with sizes that are much smaller than everyday objects we deal with, but much larger than atoms or molecules. They can thus manifest novel and potentially useful properties that can have applications in diverse areas such as electronics, defense, and healthcare. Nanotechnology is projected to be a significant segment of the US and world economy in the twenty-first century.

However, since these objects are so tiny, fabricating them takes a tremendous investment in equipment and infrastructure and in personnel to maintain and train users in these tools. Often, these are beyond the resources of academic institutions and start-ups, and support from the federal government is critical.

The Texas Nanofabrication Facility (TNF), a National Nanotechnology Coordinated Infrastructure (NNCI) site at the University of Texas at Austin, builds upon a proven NSF-supported model and will help train the next generation of engineers and scientists in this nascent field, with a strong focus on recruiting minorities and women in these STEM fields.

It will also engender start-ups in nanotechnology and will lead to advances in areas such as faster computers that would consume less energy, lasers and photonic devices for faster communication, and a dazzling array of products for the consumer market and defense applications. This effort will also have an impact on healthcare and other major federal initiatives such as developing nanosensors for the Brain Initiative, better DNA and protein sequencing tools for personalized medicine, and nanostructures for targeted drug delivery.

The TNF will facilitate breakthroughs in nanoscience and technology, with applications in nanoelectronics/photonics, green energy, and healthcare in the southwest and in the nation, by providing state-of-the-art capability in nanodevice prototyping, metrology, and nanomanufacturing. The efforts in prototyping of nanoelectronic devices will be underpinned not only by tool-training-based approaches, as in the past, but also new holistic solution-based schemes.

Coupled with device fabrication, TNF will provide cutting-edge tools in nanoscale imaging and metrology at the atomic scale. The application of nano-in-healthcare will be bolstered by the recently established Dell Medical School at the University of Texas, one of very few new medical schools in recent decades and one that is committed to reinventing medical education.

The TNF will mentor and foster start-ups in nanotechnology and provide platforms for nanomanufacturing of prototypes at the NSF Engineering Research Center in Texas. It will establish educational activities in nanotechnology especially directed at underrepresented minorities and women while making social and ethical implications (SEI) of nanotechnology an integral component of every activity.

The growing Hispanic population in the southwest makes it crucial that this segment of the population is well represented in STEM fields. The TNF is located centrally in the so-called Texas Triangle, which encompasses most of the population and the high-technology activity in Texas. This is one of eleven mega-population centers in the nation and a key hub of nanotechnology in the United States. As such, this site should have a major impact on advancing nanoscience and technology in the twenty-first century in the United States.

Midwestern Region

Midwest Nano Infrastructure Corridor (MINIC)

Investigator(s):	Stephen Campbell campb001@umn.edu (principal investigator)
	Steven Koester (co-principal investigator)
	Aaron Reinholz (co-principal investigator)
	Syed Ahmad (former co-principal investigator)
Sponsor:	University of Minnesota-Twin Cities
	200 OAK ST SE
	Minneapolis, MN 55455-2070 (612)624-5599

Abstract

Recent advances in technology allow the fabrication of very small structures with highly desirable capabilities. This enables new physical and chemical understanding (nanoscience) as well as new structures and devices that are of interest to many industries (nanotechnology). The Midwest Nano Infrastructure Corridor (MINIC) National Nanotechnology Coordinated Infrastructure (NNCI) site at the University of Minnesota will accelerate these advances by providing access to leading-edge micro- and nanofabrication capabilities for the research and development of nanoscience and technology.

The MINIC core facilities represent more than $50M in labs and equipment as well as more than 400 man-years of staff expertise. Academic researchers can use these capabilities on an equal basis with University of Minnesota faculty. Students will travel to MINIC facilities to gain valuable hands-on experience. Entrepreneurs will enjoy low-cost access to try new ideas without having to make long-term capital equipment commitments.

MINIC will support a broad spectrum of nano R&D; however, it will target researchers in two new areas: the application of two-dimensional materials and the use of nano in biology and medicine. By partnering with North Dakota State University, MINIC will also enable the packaging of nanodevices. This allows researchers to perform reliability testing and to incorporate these devices into complex electronic systems. MINIC will also reach out to underserved communities to increase their participation in this rapidly growing field. It will also support micro- and nanolaboratories at smaller schools throughout the Midwest to enable the development of nanotechnology over a broad geographic area.

MINIC will provide support to micro and nano researchers throughout the country. MINIC offers researchers access to multiple cleanrooms with a full suite of fabrication equipment including state-of-the-art electron beam lithography and extensive staff support to enable them to carry out difficult fabrication projects in a timely and cost-effective manner.

To better recruit and serve external users, MINIC will add three new process focus areas. The first will support the deposition of a broad variety of 2D thin films, beginning with graphene and the transition metal dichalcogenides. Users will be able to build devices on top of their own substrates without the low yield and variability associated with exfoliation. MINIC will also provide new modeling tools to support this area. The second focus area will be led by North Dakota State University's Packaging Center, which has long-standing expertise in the area. This will enable researchers in academia and industry to economically package nanoscale devices, including difficult applications such as RF devices, MEMS, power devices, and 3D multichips. MINIC's third focus area will support external users working in bionanotechnology by providing all the facilities and equipment needed to form nanoparticle suspensions, perform sizing and zeta potential measurements, use them to expose cell cultures in a BSL2 environment, and characterize the result with confocal and fluorescence microscopy. MINIC will also develop a novel outreach program to support nanoscience and technology labs throughout the upper Midwest.

Soft and Hybrid Nanotechnology Experimental (SHyNE)

Investigator(s):	Vinayak Dravid v-dravid@northwestern.edu (principal investigator)
	Chad Mirkin (co-principal investigator)
	Horacio Espinosa (co-principal investigator)
	Andrew Cleland (co-principal investigator)
	Samuel Stupp (co-principal investigator)
Sponsor:	Northwestern University
	1801 Maple Ave.
	Evanston, IL 60201-3149 (847)491-3003

Abstract

The Soft and Hybrid Nanotechnology Experimental (SHyNE) Resource NNCI site is a collaborative venture between Northwestern University (NU) and the University of Chicago (UC), building upon each institution's long history of transforming the frontiers of science and engineering. Soft nanostructures are typically polymeric, biological, and fluidic in nature, while hybrid represents systems comprising soft-hard interfaces.

SHyNE facilities enable broad access to an extensive fabrication, characterization, and computational infrastructure with a multifaceted and interdisciplinary approach for transformative science and enabling technologies. In addition to traditional micro-/nanofabrication tools, SHyNE provides specialized capabilities for soft materials and soft-hard hybrid nanosystems. SHyNE enhances regional capabilities by providing users with on-site and remote open access to state-of-the-art laboratories and world-class technical expertise to help solve the challenging problems in nanotechnology research and development for nontraditional areas such as the agricultural, biomedical, chemical, food, geological, and environmental, among other industries.

A critical component of the SHyNE mission is scholarly outreach through secondary and postsecondary research experience and integration with curricula at both universities, as well as societal outreach through a novel nano-journalism project in collaboration with the Medill School of Journalism. SHyNE promotes active participation of underrepresented groups, including women and minorities, in sciences, and utilizes Chicago's public museums for broader outreach.

SHyNE leverages an exceptional depth of intellectual, academic, and facility resources to provide critical infrastructure in support of research, application development, and problem-solving in nanoscience and nanotechnology and integrates this transformative approach into the societal fabric of Chicago and the greater Midwest.

SHyNE is a solution-focused, open-access collaborative initiative operating under the umbrella of NU's International Institute for Nanotechnology (IIN), in partnership with UC's Institute for Molecular Engineering (IME). SHyNE's open-access user facilities bring together broad experience and capabilities in traditional soft nanomaterials such as biological, polymeric, or fluidic systems and hybrid systems combining soft/hard materials and interfaces.

Collectively, soft and hybrid nanostructures represent remarkable scientific and technological opportunities. However, given the sub-100-nm length scale and related complexities, advanced facilities are needed to harness their full potential. Such facilities require capabilities to pattern soft/hybrid nanostructures across large areas and tools/techniques to characterize them in their pristine states. These divergent yet integrated needs are met by SHyNE, as it coordinates NU's extensive cryo-bio, characterization, and soft-nanopatterning capabilities with the state-of-the-art cleanroom fabrication and expertise at UC's Pritzker Nanofabrication Facility (PNF).

SHyNE addresses emerging needs in synthesis/assembly of soft/biological structures and integration of classical cleanroom capabilities with soft-biological structures, providing expertise and instrumentation related to the synthesis, purification, and characterization of peptides and peptide-based materials. SHyNE coordinates with Argonne National Lab facilities and leverages existing supercomputing and engineering expertise under the Center for Hierarchical Materials Design (CHiMaD) and Digital Manufacturing and Design Innovation Institute (DMDII), respectively.

An extensive array of innovative educational, industry, and societal outreach, such as nano-journalism, industry-focused workshop/symposia, and collaborations with Chicago area museums, provide for an integrated and comprehensive coverage of modern infrastructure for soft/hybrid systems for the next-generation researchers and the broader society.

Nanotechnology Corporate Partners (NCP) Program

Although also housed at Northwestern University, this is not part of NNCI; however, it has an advertised small business and industry partnership program, as found at https://www.iinano.org/nanotechnology-corporate-partners-program.

To find out more about the NCP program, please contact:

Corporate Partners Program:
Veronica Durdov
veronica.durdov@northwestern.edu
847.467.4228

Small Business Commercialization:
Kathleen Cook, Chief of Staff
k-cook@northwestern.edu
847.467.5335

Abstract
Corporate Partner Program
    In today's competitive business environment, industry is finding it necessary to cut back on research endeavors. Yet the need to stay on the leading edge of technology is undiminished. University researchers and industry have critical resources to offer each other, but sometimes need a program that can help link them together.

    The Nanotechnology Corporate Partners (NCP) Program at the IIN creates that link. NCP participants benefit from ongoing multilevel interaction with faculty, staff, and students, thereby strengthening their relationships with one of the world's leading centers for nanotechnology research. Partners gain exposure to a broad spectrum of research applications and interact with a pool of highly talented potential employees. As a member of the NCP program, your organization receives a mechanism for closer interaction between academic and corporate researchers to further mutual goals.

Small Business Commercialization
    Academic exploration in the field of nanotechnology is often driven by federal and state investment with the hope that this investment will bear fruit in the marketplace.

The IIN's Small Business Entrepreneur's Evaluation (SBEE) program provides a platform for scientists and engineers to present their newly developed technologies and receive assistance in the development of viable business plans.

Through this program, faculty members are offered the opportunity to present their marketable technology to an audience of students from the NU Kellogg School of Management, who, then, may use this as a springboard for writing a complete business plan. The success of the SBEE program is evidenced by the formulation of 20 start-up companies since the inception of the IIN who have raised over $700 million in venture capital to date.

Southern Region

The Kentucky Multi-scale Manufacturing and Nano Integration Node (MMNIN)

Investigator(s):	Kevin Walsh walsh@louisville.edu (principal investigator)
	Thomas Starr (co-principal investigator)
	Bruce Alphenaar (co-principal investigator)
	Shamus McNamara (co-principal investigator)
	J. Todd Hastings (co-principal investigator)
Sponsor:	University of Louisville Research Foundation Inc
	The Nucleus
	Louisville, KY 40202-1959 (502)852-3788

Abstract

The Kentucky Multi-scale Manufacturing and Nano Integration Node (MMNIN) is a collaboration between the University of Louisville and the University of Kentucky focused on integrating manufacturing technology over widely different length scales. With nanotechnology now integral to scientific discovery and engineering, there is a pressing need for infrastructure that supports the rapid and effective prototyping of nanoscale devices in macroscale systems.

The goal of the MMNIN is to combine micro-/nanofabrication processes with the latest in 3D additive manufacturing technology to allow researchers to explore nanotechnology solutions to real-life problems in healthcare, energy, the environment, communication, and security. In addition to having access to state-of-the-art tools and expertise, MMNIN participants will conduct novel research addressing multiscale manufacturing and integration challenges with a particular focus on the interfaces between different length scales, materials, and manufacturing processes.

The MMNIN's state-of-the-art multidisciplinary infrastructure will serve a growing user base whose home institutions include high schools, university laboratories, government facilities, start-up ventures, and Fortune 100 companies. Through new educational and seed research programs, MMNIN will offer unique opportunities to users from traditionally underrepresented regions and groups, including the Appalachian region of the United States. MMNIN's unique focus and central location (60% of the US population within a day's drive) will greatly encourage external usage. Through these efforts MMNIN seeks to transform the interfaces between the nanoscale and the human scale and impact society by rapidly providing new multiscale technological solutions.

The MMNIN will be the first open user facility nationwide with a focus on 3D micro-/nanofabrication and true multiscale integration. Users will have access to design, simulation, and fabrication resources that span the nanometer to meter scales and the expertise to effectively integrate these processes. At the nanoscale, MMNIN will provide rapid prototyping capabilities based on electron- and ion-beam-induced processes and two-photon polymerization along with the expertise to convert the prototyped structures to functional devices.

At the microscale, users will have access to a variety of unique fabrication processes including stress engineered thin-film deposition for self-programmed 2D to 3D fabrication; 128-level grayscale lithography for rapid prototyping of complex 3D structures; micro aerosol jet 3D printing using conductive, resistive, dielectric, and biological materials; and a diversity of traditional semiconductor and MEMS fabrication processes using MMNIN's new class 100 $30M, 10,000 sq. ft. cleanroom facility.

At the meso-/macroscale, MMNIN offers automated roll-to-roll manufacturing processes and the latest in additive manufacturing tools for 3D printing custom structures and enclosures using metals and/or polymers. MMNIN also offers a variety of characterization techniques ranging from transmission electron microscopy to squid magnetometry. All of these efforts involve exciting research challenges, not only on the processes themselves, but also on the integration of these processes to make reliable electrical, mechanical, optical, and fluidic interfaces between the length scales.

As a result, users will be able to create systems such as nanoscale sensors in biocompatible enclosures, artificial crystalline optical filters with high-density interconnects, and nanoelectronics that expand from the substrate to interact with the external world. Ultimately, these capabilities, combined with the MMNIN faculty expertise in multiscale manufacturing and integration, will allow users to rapidly and economically produce products and solutions addressing society's pressing challenges.

Southeastern Nanotechnology Infrastructure Corridor (SENIC)

Investigator(s):	Oliver Brand oliver.brand@ece.gatech.edu (principal investigator)
	Shyam Aravamudhan (co-principal investigator)
	Daniel Herr (co-principal investigator)
Sponsor:	Georgia Tech Research Corporation
	Office of Sponsored Programs
	Atlanta, GA 30332-0420 (404)894-4819

Abstract

Development of nanoscale materials and devices, as well as incorporation of these components into full systems, has become an important part of addressing global challenges in energy, health, and the environment. However, nanoscale science and engineering often requires the use of complex and expensive tools and facilities for the fabrication and characterization of these materials and devices. This necessitates the support of shared national resources for both basic research in academic institutions and the translation of these discoveries into commercial products by small and large enterprises.

As part of the National Nanotechnology Coordinated Infrastructure (NNCI) program, the Southeastern Nanotechnology Infrastructure Corridor (SENIC) will create a partnership between the Institute for Electronics and Nanotechnology at the Georgia Institute of Technology and the Joint School of Nanoscience and Nanoengineering, an academic collaboration between North Carolina A&T State University (NCA&T) and the University of North Carolina at Greensboro (UNCG).

This national resource will provide open access to nanofabrication and characterization facilities and tools along with expert staff support to a growing user community across the southeastern United States. The SENIC infrastructure will strengthen and accelerate discovery in nanoscience and nanoengineering, benefiting both traditional disciplines, such as electronics and materials, and newer areas, such as biomedical and environmental sciences.

In addition, because societal and economic need requires a skilled workforce trained in the tools and techniques of nanotechnology, SENIC will implement a comprehensive education and outreach program, embedded with lessons in socially and ethically responsible development and use of nanotechnology, designed to reach a broad and diverse audience of students, teachers, and the public.

With access to more than 230 nanotechnology fabrication and characterization tools, SENIC's goal is to provide a one-stop-shop approach, covering both top-down approaches using nanoscale patterning and bottom-up approaches based on nanomaterial synthesis and additive processing. A particular strength of the partnership is the ability to connect nanomaterials and devices to full packaged systems. This helps transition nanoscale research achievements more quickly into high-impact applications in biomedical/health, energy, communication, smart transportation, textiles, and smart agriculture.

SENIC will operate with an interdisciplinary culture where engineers, scientists, physicians, educators, policy experts, and economic development professionals work together with shared access to facilities and tools and a deep understanding of industry opportunities and societal challenges to promote the accelerated translation of invention into innovation. Furthermore, the SENIC partners will work with undergraduate and graduate students, as well as partner with 2-year technical colleges, to produce science and engineering professionals from diverse backgrounds who are ready to meet the global workforce demands of the twenty-first century.

In tandem, the public outreach with hands-on classroom activities as well as interactive facility tours will encourage K-12 students to participate in the STEM pipeline and will help create an informed citizenry that supports the safe development of nanotechnology. Closely coupled with its education program, SENIC will have a social and ethical implications program that educates on the challenges associated with the expectations that nanoscale science and engineering will contribute to the solution of societal, environmental, and economic problems, while anticipating and avoiding potential negative consequences.

Eastern Region

Cornell Nanoscale Science and Technology Facility (CNF)

Investigator(s):	Daniel Ralph (former principal investigator)
Sponsor:	Cornell University 373 Pine Tree Road Ithaca, NY 14850-2820 (607)255-5014

Abstract
The Cornell Nanoscale Science and Technology Facility (CNF) will provide the nation's researchers with rapid, affordable, hands-on shared access to advanced nanofabrication tools and associated staff expertise that are too expensive for individual universities or small companies to operate and maintain.

Under this National Nanotechnology Coordinated Infrastructure (NNCI) site award, hundreds of engineers and scientists nationwide, from throughout academia, industry, and government, will utilize CNF's unique toolset and technical staff. The new research and technology development that the CNF makes possible will transform many fields of engineering and science, spanning sensor and actuator arrays for probing how the brain works; improved photovoltaics, batteries, and fuel cells for economical renewable energy; new types of electronic devices that surmount limitations of silicon; fabrication of living tissues and organs; distributed measurement networks for geosciences; microbiome characterization and manipulation; on-chip signal processing with light; precision agriculture using new sensors; low-cost medical diagnoses; and improved quantum devices for utilizing entanglement, to name just a few.

The CNF will also organize education and outreach programs targeting a wide range of researchers, undergraduates, and K-12 students and their families, with the broader goals of providing a hands-on research education to a new generation of diverse engineering and science students, facilitating the commercialization of nanotechnology for societal benefit, extending the benefits of nanofabrication to less-traditional areas of research, and interesting more young students in technology and science.

The unique nanofabrication capabilities that the CNF will make available to the nation's researchers include world-leading electron beam lithography, advanced optical lithography, dedicated facilities for soft lithography, and direct-write tools for rapid prototype development, along with the flexibility to accommodate diverse projects through the ability to deposit and etch a very wide variety of materials.

In addition, CNF's experienced, expert technical staff will be solely dedicated to user support. Engineers and scientists from throughout academia, industry, and government will utilize CNF's resources to fabricate structures and systems ranging from the centimeter scale down to the nanometer scale. The new research that the CNF makes possible will advance research and development across many fields, spanning electronics, optics, magnetics, mechanical devices, thermal and energy systems, electrochemical devices, fluidics, and the life sciences and bioengineering.

Educational and outreach programs are designed to benefit several target audiences. CNF will teach new nanotechnology researchers to quickly operate at the research frontier through hands-on training on state-of-the-art tools, in-depth minicourses, and subject area workshops. Experienced nanotechnology researchers will benefit from advanced technical workshops in partnership with leading tool vendors. Undergraduate students will participate in blended courses with regional universities and unique summer research experiences. Young K-12 students and their families will be encouraged to become excited about technology and science through CNF's highly-popular Nanooze science magazine and a new partnership with 4-H.

Virginia Tech National Center for Earth and Environmental Nanotechnology Infrastructure (VT NCE2NI)

Investigator(s):	Michael Hochella hochella@vt.edu (principal investigator) Frederick Marc Michel (co-principal investigator) Amy Pruden (co-principal investigator) Linsey Marr (co-principal investigator) Peter Vikesland (co-principal investigator)
Sponsor:	Virginia Polytechnic Institute and State University Sponsored Programs 0170 BLACKSBURG, VA 24061-0001 (540)231-5281

Abstract

The scientific and engineering investigations and understanding of ultrasmall objects, known as nanomaterials, is not only revolutionizing critical fields such as medicine, personal electronics, and national security, but it is also sharpening the understanding of how the Earth works. The air one breaths, the soil in which crops are planted, the metals from minerals that build industries, and the contaminants that can profoundly harm one (from arsenic to dangerous bacteria) are all related to and/or influenced by the vast store of nanomaterials that make up key portions of the planet.

These nanomaterials must be studied and understood, despite great difficulty due to their minute size, to safely and efficiently clean the air, purify water, and allow the responsible use of Earth's vast store of life-sustaining resources. The Virginia Tech National Center for Earth and Environmental Nanotechnology Infrastructure

(VT NCE2NI) will greatly accelerate the progress that Earth scientists and engineers have made in studying, understanding, explaining, and utilizing Earth for the well-being of all. In addition, new types of environmental sensors and detectors based on rapidly emerging nanotechnologies, for example, to detect harmful living and nonliving contaminants in air, water, and soil, will clearly be a major benefit to society. VT NCE2NI users will consist of far more than professional academics and their advanced research students. Users will also come from private and publicly held companies, as well as students from high schools, community colleges, liberal arts colleges, and key minority-serving universities.

VT NCE2NI provides an NNCI site to specifically support researchers who work with nanoscience- and nanotechnology-related aspects of the Earth and environmental sciences/engineering at local, regional, and global scales, including the land, atmospheric, water, and biological components of these fields. The national presence of VT NCE2NI is significantly enhanced by a close partnership with the Environmental Molecular Sciences Laboratory (EMSL) at Pacific Northwest National Laboratory (PNNL).

NNCI geo- and environmental science/engineering users have access to both the Virginia Tech and EMSL/PNNL sites depending on specific technical needs and geographic considerations. VT NCE2NI consists of (i) the 15,000 sq. ft. Nanoscale Characterization and Fabrication Laboratory (NCFL) that houses a broad array of high-end, state-of-the-art electron-, ion-, and X-ray-based characterization tools and sample preparation laboratories, as well as meeting space and ample office space for visitors and (ii) the 6300 sq. ft. Virginia Tech Center for Sustainable Nanotechnology (VT SuN) which contains extensive nanomaterial synthesis facilities and knowhow (in aqueous, soil/solid media, and atmospheric environments), characterization tools, and experimentation/reactor systems, plus meeting rooms and additional office space for visitors.

VT NCE2NI also provides broader impact initiatives including substantial funding for students from key minority-serving institutions and outreach programs with community colleges and four-year liberal arts colleges. The overall contribution of this NNCI site will be to accelerate the growth of a field that is revolutionizing the understanding of several broad aspects of Earth and environmental sciences and engineering.

North Carolina Research Triangle Nanotechnology Network (RTNN)

Investigator(s):	Jacob Jones jacobjones@ncsu.edu (principal investigator)
	James Cahoon (co-principal investigator)
	David Berube (co-principal investigator)
	Mark Wiesner (co-principal investigator)
	Nan Marie Jokerst (co-principal investigator)

Sponsor:	North Carolina State University CAMPUS BOX 7514 RALEIGH, NC 27695-7514 (919)515-2444

Abstract

The Research Triangle Nanotechnology Network (RTNN) enables innovation and commercialization of new promising nanotechnologies and enables public education for the United States by providing technical leadership and open access to comprehensive and unique nanotechnology laboratories, equipment, and research expertise. The RTNN is anchored by three major research universities (North Carolina State University, Duke University, and University of North Carolina at Chapel Hill) that are clustered near one of the nation's major nanoscience and nanobiotechnology regional economies.

The RTNN focuses on pioneering, studying, and refining innovative methods to catalyze both traditional and emerging nanotechnology research areas, including those from biology, biomedical engineering, textile engineering, environmental engineering, agriculture, soil science, forest biomaterials, and plant and microbial biology. Since the barriers of distance, cost, and awareness often prevent facility usage by both traditional and nontraditional users, the RTNN will surmount these barriers using a variety of innovative programs.

The RTNN further leads research on the societal and ethical implications of nanotechnology, including issues of environmental health, safety, ethics, and equity, through a social and ethical implications of nanotechnology (SEIN) component that also assesses innovative program success. The RTNN will create a nanotechnology innovation ecosystem that spans grades 7–12, community colleges, universities, and industry. By translating program successes across the nation, the RTNN will become a national focal point for innovation and will serve as a guide for nanotechnology innovation ecosystems across the nation.

The RTNN brings together comprehensive shared user facilities and complementary faculty research programs at three major research universities. These resources will be used to dramatically increase the national impact of state-of-the-art fabrication and characterization facilities and research expertise in nanoscience and nanotechnology. RTNN technical capabilities span nanofabrication and nanocharacterization of traditional hard, dry materials (i.e., 2D and 3D nanomaterials, metamaterials, photonics, and heterogeneous integration) and emerging soft, wet materials (i.e., tissue, textile, plant, and animal nanomaterials).

Specific areas of capability include the environmental assessment of nanotechnology, atomic layer deposition, flexible integrated systems, and fluidic systems. The RTNN will enable emerging research areas by adding additional process flows and tools throughout the project that enable new ways of integrating and interfacing the nanoscale with the human scale. The RTNN will expand shared facilities usage by creating and assessing innovative programs and disseminating these programs across the nation.

These programs include graduate student peer-to-peer distance using Internet networking, summer undergraduate research internships with follow-on outreach to the student's school of origin, public engagement leveraging large-scale Internet courses, and outreach to grades K-12. A specific emphasis is on engaging users from underserved user groups, including women, minorities, and people who do not typically access shared university facilities. The aim of the RTNN is to create a comprehensive, integrated nanotechnology ecosystem that will provide a pipeline of STEM students for a strong, vibrant, and entrepreneurial next-generation workforce.

East Coast Region

The Center for Nanoscale System (CNS) at Harvard University

Investigator(s):	Robert Westervelt westervelt@seas.harvard.edu (principal investigator)
	William Wilson (co-principal investigator)
Sponsor:	Harvard University
	1033 Massachusetts Ave
	Cambridge, MA 02138-5369 (617)495-5501

Abstract

The goal of the Center for Nanoscale Systems (CNS) at Harvard University is to provide outstanding facilities and expertise to make, image, and understand nanoscale structures and systems. CNS provides a collaborative, multidisciplinary research environment that allows researchers from academia and industry to study and develop new structures, devices, systems, and technologies in fields ranging from biomedicine to nanoscale electronics and photonics.

CNS offers tools for nanofabrication, electron microscopy, and characterization of nanoscale systems, with technical expertise and assistance provided by its staff. CNS is one of the most active nanofabrication and imaging facilities in the world with more than 1500 users, and it is an important part of the high-technology boom in the northeast. With its diverse user base, well-established infrastructure, and outstanding facilities, CNS is well placed to continue as a technology leader.

In addition, CNS plays a key role training the nation's next generation of scientists and engineers. It has an established Research Experiences for Undergraduates (REU) program, as well as an annual summer nanotechnology seminar series. A new CNS Scholars Program will bring in underrepresented researchers, and an internship program will train US veterans in nanotechnology. As part of the NNCI, CNS will help make the transition from research on nanoscale devices to complex nanosystems engineering.

Since its creation in 2001, CNS has become a key nanotechnology resource for the nation. As part of the previous NNIN, CNS developed diverse and versatile facilities including multi-length-scale optical and electron beam lithography, focused ion beam (FIB) and reactive ion etch (RIE) systems to shape structures, and soft lithography expertise to enable fabrication of a wide variety of microfluidic systems. These tools allow users to push the frontiers of nanoscale electronics and photonics using nontraditional materials, and they enable the development of sensor systems for biomedicine.

CNS researchers pursue advanced topics including plasmonics, diamond photonics, nanoscale sensors, and atomic-layer devices. CNS has an outstanding suite of imaging and characterization tools including an aberration-corrected STEM, a high-resolution TEM, a CryoTEM, and an atom probe for 3D tomography, as well as scanned probe microscopes, and linear and nonlinear optical microscopes. Its characterization tools permit detailed analysis and assessment of materials, components, and systems, providing researchers with a comprehensive platform for nanotechnology research.

CNS focuses on the core missions of the National Nanotechnology Initiative (NNI): advancing world-class nanotechnology research, fostering the transfer of new technologies into products for commercial and public benefit, developing and sustaining educational resources to develop a skilled nanotechnology workforce, and supporting the evolving infrastructure and advanced tools needed to support excellence in nanotechnology research and development.

Mid-Atlantic Nanotechnology Hub (MANTH) for Research, Education, and Innovation

Investigator(s):	Mark Allen mallen@seas.upenn.edu (principal investigator)
	Cherie Kagan (co-principal investigator)
	Kevin Turner (co-principal investigator)
Sponsor:	University of Pennsylvania
	Research Services
	Philadelphia, PA 19104-6205 (215)898-7293

Abstract

Nanotechnology, the exploitation of the science of the very small, has the potential to revolutionize lives—a few examples include faster electronics, smaller biomedical implants, better batteries, materials with high strength or the ability to self-clean, and small machines that can sense the physical world. Often, nanotechnology development requires access to a large fabrication and characterization infrastructure and the related scientific expertise to allow users to build and measure such nanomaterials and nanodevices.

This project establishes a Mid-Atlantic Nanotechnology Hub (MANTH) for research, education, and innovation at the University of Pennsylvania as an NNCI site. This site will allow users in the mid-Atlantic region, the nation's fifth largest economic area, to access the Singh Center for Nanotechnology, where they can perform nanofabrication and measurement tasks and interact with nanotechnology experts. The Singh Center is located at the University of Pennsylvania in downtown Philadelphia and is highly accessible to over 100 regional academic institutions and the industry-rich mid-Atlantic region.

The Singh Center will also host education programs to introduce high school students, college undergraduates, and the Philadelphia area community to nanotechnology. These potential future nanotechnologists will have the opportunity to participate in Nano Day and Summer Research Experiences for Undergraduates programs held on-site, view nanotechnology contributions to community outreach programs such as Philly Materials Day and the Philadelphia Science Festival, and participate in workforce training activities for nanotechnology technicians carried out in partnership with the Community College of Philadelphia.

MANTH will enable access to leading-edge research and development facilities and expertise for academic, government, and industry researchers conducting activities within all disciplines of nanoscale science, engineering, and technology. Examples of its capabilities include: electron beam, photo-, imprint, and soft lithographies, material deposition and etching, multiscale 3D printing, laser micromachining, electron and scanning probe microscopy, tip-based nanofabrication, and ion and electron beam milling.

This NNCI site will foster intellectual collaboration by assisting users in addressing their nanoscience and application needs, providing a forum for intellectual exchange between academic and industry users, and developing new fabrication processes that not only contribute to the users' end application but also further the fields of nanofabrication and nanomanufacturing. Users will also benefit from interaction with Penn faculty and staff, possessing significant expertise in nanofabrication; flexible/stretchable nanodevices; MEMS; microfluidics, photonics, and electronics; fabrication and exploitation of nanostructured soft materials systems; multiple low-dimensional materials; characterization and integration of beyond-silicon material systems; and investigations at the nano-bio interface, from medical devices to enabling the understanding of biological systems.

In addition to serving existing researchers, MANTH will engage potential future nanotechnologists through on-site research reviews, summer research experiences, and outreach to local Philadelphia high schools. By acting as a catalyst for growth of nanotechnology in this region, significant opportunities for nanotechnologists at all levels will be created, from technicians through a workforce development partnership with the Community College of Philadelphia to PhD researchers.

Chapter 14
Meet the Lab Directors

Each of the 16 NNCI facilities has its own leadership team. There is usually an investigator(s) who is responsible for interfacing with NNCI, a sponsor which is usually a university or facility at a university that provides the space and resources for the lab, and then the lab director(s) who runs the lab operations on a day-to-day basis. These people are your allies, and their mission is for you to have a successful experience at their facility, so get to know them! Below are interviews with two lab directors, Dr. Michael Khbeis at the University of Washington and Dr. Jiangdong "JD" Deng at Harvard University.

Michael Khbeis

Dr. Michael Khbeis was the director of the Washington Nanofabrication Facility (WNF) at the University of Washington. He earned his doctorate and has 20 years of experience in the semiconductor industry. His doctoral research was on energy harvesting with MEMS, and his research interests are in advanced packaging of RF electronics and circuits. After working for the Department of Defense in a MEMS facility for 10 years, he joined the WNF 7 years ago. He recently left the WNF and is replaced by a new director, Dr. Maria Huffman from Lund University.

Michael says their commercial/industry interactions vary. An NDA is signed before starting. Mostly, they conduct a design review, find out what the client's needs are, and determine who is going to do the work—staff at the lab, you, or engineers you are providing. The design review is used to brainstorm a fabrication plan and sometimes to provide a cost estimate (which can be challenging). For projects that are very complex, they will not do on a contractual basis, because they simply cannot commit to that amount of time. For those projects, the client

© Springer Nature Switzerland AG 2019
D. Munro, *DIY MEMS*, https://doi.org/10.1007/978-3-030-33073-6_14

must provide their own employees. When everything is agreed, contracts are signed, purchase orders to bill against are submitted, and some kind of "not to exceed" amount is stipulated.

Clients interface with the lab in different ways. Some people fly in, do their project, and leave. Others have full-time people working there in the lab. Those people tend to have long-term desk rentals, but others use "hoteling areas" with shared desk spaces and lockers (there are 100+ lockers to store non-cleanroom or excess items and 100 cleanroom boxes). Right now they are renovating, so soon there will be many more informal and formal meeting spaces.

They are adding capability to their facility as well, including their most popular basics (CVD, isolation for dielectrics, and lithography—laser, electron beam, photo, and contact), as well as the newest technologies (advanced DRIE, nanoscale 3D printer, vapor phase etchers with HF, and atomic layer deposition).

To contact WNF, he recommends you either directly email or make a phone call. You can also check out their website or go through NNCI. Many of their clients get connected through word of mouth, too. Most workers tend to live locally, but there are people who rent corporate housing. [*As a side note, I stayed at a nearby (walking distance) pension hotel (breakfast included) with bathrooms down the hall, and that worked well. Over the summer, the University of Washington also rents out its residence halls like hotel rooms. They have beddings, towels, soaps, and cleaning service, and there is the added benefit of dining hall meal service, if desired, although restaurants abound nearby and on campus.*]

All new people are put through a training regimen in order to safely use the lab, including equipment-specific training as needed on demand. Training is done as a two-part process—an initial introduction followed by checkout training where the new user demonstrates their proficiency operating the equipment while being supervised by staff. Some equipment, such as SEM and E-beam lithography, take additional training that is conducted over multiple sessions.

The lab is happy to support people with anything they need. Users come in all skill levels, from complete novices to industry veterans, so their goal is to "meet them where they are at" and tailor their approach to the client, whatever that may mean—process support, anecdotal advice, short loop experimentation, help with purchasing, etc. No matter who you are, they treat everyone the same—with respect, courtesy, and professionalism.

Right now, they are trying to do a better job of cross-engagement between NNCI sites. Several ideas are in the works, including allocating some resources to share ideas on an email forum, providing reciprocal advice, and teaching/sharing use of equipment. Right now, even if WNF needs a piece of equipment available at another site, they will use a subcontractor (like Rogue Valley) to do what they need. This is a challenge they would like to solve, but it is still up in the air on how to keep projects within NNCI when the equipment exists.

Michael has been to other sites, and he says that for the most part, they operate the same, although the University of Washington is more industry-centric than most facilities. The equipment has about 90% of the same types of capabilities, with 10%

being niche, specialized equipment. [*If you have a niche need, it is still best to begin with your regional NNCI lab, however, and then get advice on how to bridge over to the specialized equipment through NNCI. You could be part of the solution on this type of interfacility collaboration!*]

NNCI is set up as primary and satellite sites. The primary sites are geographically partnered with satellite sites in order to be inclusive and also to have a better network across a region, rather than just in nodes of high tech. For example, WNF is a full capability lab, and its satellite at Oregon State University (OSU) has specialized materials and niche capabilities.

There *are* open use facilities outside of the NNCI consortium. He mentioned the University of Michigan as a good option. It is not in the network anymore, but still allows commercial companies to use the facilities. The University of Wisconsin-Madison also has a facility, he said, and in fact, almost any major school has a microfabrication facility—whether they engage with industry, are friendly about IP, etc., however, varies dramatically. There is no listing of these facilities, but if you want more information, most lab managers attend the annual UGIM conference (Users Group on Innovation through Manufacturing in MEMS/Nano). It is held in June at a university each year, and you are welcome to attend. You will not see a lot of commercial labs, just institutions and government labs, but it is a good place to begin networking if an NNCI facility will not work for your needs (technologically, geographically, or something else). UGIM is not a technical conference— it is for operational information, best practices, and information sharing.

JD Deng

JD Deng has been the lab manager at Harvard's Center for Nanoscale Systems (CNS) since 2010. He joined Harvard as a physicist in 2004, and he has 14 years of experience in MEMS and nanotechnology.

JD began his career industry doing R&D on nanoimprinting for telecom optical components. His background is in physics with a speciality in optics, and he earned his doctorate in China. One of his PhD is in scanning optical microscopes, but his areas of research interest are micro-optical fiber sensors, chemical sensors, and acoustic sensors, as well as nanofabrication technology—lithography, scanning probe microscopy, 3D laser printing technology, and nanocharacterization. He has developed humidity sensors for buildings, photonics devices, and medical sensors.

Harvard's CNS is a shared center open to outside users. It was built in 1999 by several key professors, with Harvard University funding $200 M for the building and startup team. They moved in in 2007 and now have 1600 users—the largest in the NNCI network. Usage was 190,000 hours last year! They have divided their facility into two technical ports—nanofabrication and imaging.

On the nanofabrication side, which JD manages, they have 10,000 sq. ft., with both Class 100 and 1000 cleanrooms. In a typical month, JD's side averages about 1000 users putting in 9000–10,000 hours. One unique feature about CNS is that they operate their cleanrooms differently, allowing not only silicon, but any kind of material—silicon dioxide, carbon nanotubes, other 2D materials, diamond, and flexible polymers, to name a few.

Currently, CNS does not have many industry users, maybe 10–15% (around 200 people). Academia accounts for way more than half of their users, and 60% are internal to Harvard. An additional 10–20% are from MIT, even though they have a cleanroom fabrication facility, because the CNS fab is better and more flexible (MIT is Si-only). The remaining 15–20% are other universities—Yale, Cornell, and Princeton.

Most distance commuters (1–2-hour drive away) will come work daily. More remote commuters (like Princeton) will come and stay for 1–2 weeks. International users will come for 1–2 months and establish an internal collaboration with a Harvard person. Funding is always an issue when trying to establish a collaboration, but he said it still makes sense if you are not able to work on-site.

CNS has an Incubator Program, where small companies can join together and make *one* account to share resources. Or you can come separately and pay for just what you need. They are happy to work with you to find the best option for your needs.

There is no office space at CNS, which is always a big issue. They have open tables and some storage, as well as some open use desks in the hallway (about 20–30) and a few computer rooms for designing. He reassured me that "you can always find a place to sit." There is storage in the gowning area, called the "tool-box," and outside the cleanroom, you can rent a cabinet. It is not huge, but adequate for most users.

People have to find their own housing (private housing is often available for rent near campus), although there is a Research Experience for Undergraduates (REU) program for students that includes housing in the summer. So, if you hire a student to work at the lab, they can obtain housing. CNS has a parking garage that you can get a permit to use that is a few blocks away.

Getting started is easy to set up. Just go to their website and do a user enrollment form—name, PI (principal investigator), and project idea. Basically, he wants to determine what kind of project you want to do, so JD reviews your form personally to see if it is feasible at CNS. Then, you are assigned a temporary user account.

CNS uses a standard cleanroom protocol for clothing—no open shoes, and they do not recommend shorts or short sleeves. You are allowed to bring in phones and laptops if you wipe them down.

Then, you go into training. Safety training is first, then cleanroom orientation and access, and then tool equipment training (there are over 50 trainings per week available online). Within 1 week of joining, you can work on at least one tool. The amount of training depends on the tool, your experience, and the usual parameters. E-beam lithography takes several classes. The lab charges a small fee for all training to cover their costs.

You can also work in "Assisted Mode," where you receive some training, but still need assistance with your design, fabrication processes, etc. In Assisted Mode, you can schedule an assistant to help you for 1–2 hours on any tool, process, or design for an extra charge. [*This can significantly speed up the process and improve the quality of your MEMS devices, so it may prove to be a cheaper option in the long run.*]

At CNS, you can also send in your design and have the staff build it. This is called "Remote Assistant Mode" and is uncommon because everyone is so busy. The lab definitely prefers users to come in and use the tool by themselves.

Finally, there is "Collaboration Mode," which is a high level of involvement, usually with a faculty member at Harvard, on the process of design, etc. If you choose this option, you have to be careful about IP, as the university will try to latch onto anything one of their faculty develops.

CNS has very friendly, nice staff, so you will receive lots of free advice. JD also holds biweekly discussion sessions in his office where you can get free design or process advice. The goal of CNS is for you to be successful.

CNS has more specialty equipment than most NNCI facilities. It is a comprehensive fab facility with almost all types of equipment. One of their strengths is in patterning and lithography, and they can have resolutions from a few nanometers to hundreds of micrometers all the way up to millimeters; thus, this equipment receives very heavy use. Their e-beam lithography is the best in the world, and they can do 2D and 3D printed patterning at sub-micrometer resolution. They also have laser direct write systems and mask writers.

Their etching tools are also very good and new, including a diamond dedicated carbon process, an oxide etcher for silicon for MEMS, and DRIE for optical MEMS. In addition, they have dedicated etchers for special materials and for metals.

They also have an AFM, over 30 sputtering materials, PCVD, and a host of great metrology equipment. They can even accommodate bio-/flexible materials—there is a dedicated room for their fab.

CNS is well staffed with 14 staff on the nanofab side (combination of scientists, engineers, and technicians) and 10 on the imaging side (to operate 5 SEM, 5 TEM, μCT, 6 AFM). People tend to stay employed at the lab for a long time. Recently, he noted they lost one person to a local company, but overall they are very stable.

Their administrator, Jim Rainoff, will work with you on enrollment and setting up your project. They also have a financial person. When you enroll and become a CNS user, you will need a purchase order (PO) account number. Every month, you will get an invoice, and the charge depends on your tool usage (each tool has different rate). There is an industry rate and an academic rate, which are posted online.

At CNS, they *do not sign NDAs*. There are just so many people; it is impossible to remember. They do not take ownership of IP, ever, but users must protect their own IP. If you need lots of help with your project, they recommend going into Collaboration Mode. They have had no issues so far. Industry clients soon get comfortable with the rules after using the lab and find they like the collaborative atmosphere where they can get idea from other users and facility staff. He says, in general, "sharing benefits everyone."

Variability Between Labs

In these two examples, you can see there is variability between the labs, but on the whole, the theme is the same—they have great equipment and technical staff at your disposal, and they want you to come and use it all! So, what are you waiting for?

Chapter 15
Collaborating with a Facility

Finding the Best Lab "Fit"

Finding the best fit depends on your project's needs and how you intend to interact with the lab. In general, I would expect the lab closest to you regionally is the best choice, as you can be more actively involved and travel becomes more feasible. However, if you are looking for an academic collaborator, you will want to use the facility closest to them.

Unless you have a very niche need (such as soft MEMS, which is done at only one facility), any of the labs will likely have the equipment you need for most or all of your project. Those skills or equipment they do not have can be readily outsourced to a vendor, or potentially to another NNCI facility (if you would like to be a pioneer and help establish that pathway).

Below, I have outlined the basic steps for working with a facility and provide advice on different ways of collaborating with a lab.

Basic Steps for Working with a Facility

All of the facilities follow approximately the same procedure for new users to get involved. The steps outlined below were pulled mainly from Cornell University's website and an interview with Dr. Michael Khbeis, at the University of Washington Nanofabrication Facility (WNF), but the advice is consistent with what other facilities recommend.

1. The first step is to contact the user program manager for the facility. I would recommend initiating contact via email and arranging for a teleconference with the user program manager. If feasible, it also makes sense to visit the facility for an informational visit. This would allow for introductions and a tour. There is usually a fee per person for a tour, as it will require gowning up and entering the cleanroom.

© Springer Nature Switzerland AG 2019
D. Munro, *DIY MEMS*, https://doi.org/10.1007/978-3-030-33073-6_15

2. You will then write up a brief project proposal of 1–2 pages in length and an application to use the facility. If confidentiality is a concern, this can be preceded by establishing nondisclosure agreements.
3. It is highly recommended to have a design review and project feasibility meeting at the facility. You will bring a more detailed presentation of your project, and the facility will arrange to have all their subject matter experts present to formulate a design and fabrication plan to meet your needs. This will also create a preliminary equipment and lab use training plan for you, if you intend to do the fabrication yourself.
4. To get established at the lab, you will need to obtain a user ID, access cards, and lab safety training. This can take a couple of weeks to complete, so plan ahead and be proactive about doing the training, tests, and paperwork required.
5. You will also need to set up a purchase order that the facility can bill against.
6. If you are going to have someone travel to work on-site, most facilities can arrange for campus housing (especially during the summer). The rest of travel and lodging arrangements are up to you, but the facilities can provide advice to assist you.
7. When you arrive at the facility, you will be assigned a staff person to mentor your project, particularly in the early phases. They will help arrange for your training, and they will be your main contact throughout the project. They will also help you define what you need, create a process flow document, purchase supplies, and subcontract with outside vendors for certain aspects of your project.
8. If you are a short-term user, you will likely work in a shared office space, which is generally first come, first serve. For longer-term users, you may have a designated space. Either way, expect to bring your laptop with you each day. Other supplies and equipment can be stored in lockers and/or cleanroom lockers. There are rules about what you must wear under your cleanroom suit, usually a long-sleeved shirt, long pants, and shoes that fully cover your feet. Hair must be tied back.

Remember to be flexible in your approach. As you learn more about your design, the facility, and the possible fabrication options, your design and process flow document is likely to change.

Initial Design Review

The first step is often a discussion with the NNCI facility's director. The principal investigator (PI) and their contact information is listed for each facility on the NNCI site, along with their NSF Award Abstract, which gives a nice summary of their facility's mission, capabilities, and additional networking information.

For me, this initial discussion included nondisclosure agreements, and we met in a conference room at the University of Washington, where we had a design review of my project. Experts from the facility weighed in with their ideas, and we developed a training plan for my use of the lab. Over the next several weeks, I was able to fill out a user agreement; arrange short-term, summer housing at the university; researched a convenient means of commuting on public transit; and proceeded to work my way through all of the training I needed.

Training began with safety protocols then stepped through gowning procedures for the cleanroom, use of microscopes, and how to clean and spin coat a film on a wafer. Rather than overwhelm me with too much information at once, the lab staff demonstrated each piece of equipment I would need one at a time, allowing me to practice and then use that equipment before having me read the training materials for the next step. It took about 2 months to complete all the training I needed, and then I was free to use the equipment. I spent about 2 years working at this facility, a few days per week during the summer and once or twice a month during the academic year, and I was able to fabricate two complete design iterations of my device.

Your use of a facility will depend on your needs. It is not necessary to go into the cleanroom yourself and do the fabrication. Each facility has expert staff that can do the design research and fabrication for you. In addition, most of these facilities are housed at a university and you can sponsor a student research assistant to pursue your design. Finally, some facilities, like the University of Washington, rent desk space to employees from private companies at their lab. These employees work for industry (maybe even you?) and conduct research in the lab full time.

Establishing a Working Agreement

All NNCI facilities will want you to establish a working agreement. The following is taken from WNF's user manual, but it is representative of what all the facilities require. Labs are busy places with lots of expensive equipment and hazards, so it is important that you come in with an open mind-set about all the rules and guidelines that have been established for everyone's benefits.

Using the Facility

Users are classified as either UW (internal students, faculty, staff, or business unit with a UW budget number) or external (non-UW). External users are further categorized as academic (other academic institutions, US governments and agencies, and certified nonprofit organizations) or industrial (all other for-profit organizations). As a user facility, the most common paradigm is for individuals to process their devices in person (on-site users). An alternative approach is contract facility staff to perform limited scope fabrication processes (remote users).

On-Site Users

While working at the WNF, you will learn a variety of processes and gain valuable skills. After a lab orientation and wet bench training, you will sign up for equipment training as needed. We advise finding a mentor or consulting with staff to ensure proper cleanroom technique, especially if you plan to repeat or expand upon an established process.

Remote Users

WNF staff engineers are available to conduct limited-scope process work on a best effort, time, and materials basis for remote users. Due to the experimental nature of most contract processes, we cannot provide product guarantees, but will work closely with clients to determine project feasibility, to provide cost estimates, and attempt to obtain mutually satisfactory results. Remote users will be assessed a fee for sample shipping.

Gaining Initial Approval

WNF's website will direct you through the induction process. In short, the process involves determining your role (UW or external, remote, etc.) and then entering contact and billing information and a project proposal. After reading and understanding this user manual, you must complete and return the appropriate Facility Use Agreement, either UW or non-UW. Optionally, you can complete a nondisclosure agreement (NDA). Lastly, you will also be required to complete a variety of online and in-person training, explained in more detail on the website.

Reports and Acknowledgments

As a provision of the National Nanotechnology Coordinated Infrastructure (NNCI) program, we are required to submit an annual report on active projects. Occasionally, WNF staff may request input in compiling publication lists and highlighting research in our labs. Additionally, you are required to acknowledge work conducted at the WNF in your publications and presentations. A suggested acknowledgment is: "Part of this work was conducted at the Washington Nanofabrication Facility, a member of the NSF National Nanotechnology Coordinated Infrastructure." Your cooperation in response to these requests is mandated by federal funding sources and greatly appreciated.

Billing

The WNF is a nonprofit business unit within the University of Washington that charges time and materials on a cost reimbursement basis with monthly invoices. Academic pricing is achieved through UW, state, and federal grants and subsidies. Industrial rates are set for cost recovery. UW Management Accounting and Analysis (MAA) mandates an annual rate proposal to ensure cost recovery. Rates are a function of cost of operation (labor and materials) and utilization; if the user base grows and utilization increases, costs decrease.

Furthermore, the rate structure falls into four tiers: basic, low, mid, and high. Users of the WNF are eligible to use characterization capabilities under a single purchase order, but will receive a separate monthly invoice from each facility.

Due to the nature of work at the facility, it is not possible to issue binding quotations for projects. Many projects have unanticipated changes in scope and processes based on experimental results, so a "not to exceed" purchase order is recommended to account for possible changes with minimum logistical overhead.

Consulting

Staff engineers are readily available for design reviews. These consulting sessions are a venue to discuss project concepts, process flows, or specific technical issues. Current or potential users are encouraged to hold discussions with staff on a regular basis to work through processes and to troubleshoot problems. Independent design reviews can be scheduled as needed by contacting the lab director or engineering staff.

User and Staff Meetings

User meetings are held every Monday afternoon. These are open forums for registered lab users to discuss issues pertaining to the laboratory, instrumentation, and processing. To ensure a timely response to issues and concerns, the WNF staff meets when the user meeting is adjourned. If you cannot attend the weekly meeting, please contact the lab director or engineering staff for alternative solutions.

Training

The following was excerpted from the University of Washington's health and safety documentation and Accident Prevention Plan (APP), available online through WNF's website.

New Employee Health and Safety Orientation

Departments must ensure that all new UW employees, including those who are temporary or part-time, undergo a health and safety orientation. The health and safety orientation must cover the following topics:

1. Information on how to find and utilize this APP and any supplemental department-specific health and safety policies.
2. Reporting procedures for fire, police, or medical emergencies.
3. Building evacuation procedures during an emergency.
4. Location of fire alarm pull-stations and fire extinguishers.
5. Procedures for reporting all accidents and incidents to their supervisor and completing a written online report using the Online Accident Reporting System.
6. Procedures for reporting unsafe conditions or acts to a supervisor, and, when possible, taking action to address unsafe conditions.
7. Location of first aid kits.
8. Information about chemicals or hazardous materials used in an employee's work environment, including how to identify them and where to locate the safety data sheets for hazard information.
9. Identification and explanation of all warning signs and labels used in their work area.
10. Use and care of any required personal protective equipment (PPE).
11. Description of any work-related safety training course(s) the employee is required to complete.

A safety orientation checklist for supervisors to onboard new employees can be found on the EH&S (Environmental Health and Safety) website.

Preparatory Work

Before getting started with an NNCI facility, you need to do some prep work. As an individual, or with your company, you should develop a plan to support your MEMS device project, asking yourself all the usual questions:

- Who is going to work on this?
- When does the project need to be done?
- What are the key milestones?
- How much can we spend on this?
- Which facility do we want to use?
- Who will be the liaison with the facility?

Once those questions are answered, you need to invest some time getting situated—How will you travel to the lab? Where will you stay? How often will you work at the lab? Do you need ground transportation once you're there? How will you feed yourself? [*Personal note, eating out for every meal for weeks on end is terrible. Find a local grocery store and choose a place to live with at least a refrigerator and a microwave. Having a hot breakfast and coffee available is also a plus!*]

After you have thought through the logistics, make contact. All of the NNCI facilities really want to grow their industrial user base, so they are eager to work with you.

Scheduling

The key to success for working at a facility is scheduling. You need to plan ahead. Below are the key things to remember.

Design Reviews

One of your first meetings will be a comprehensive design review. All of the essential personnel that might be of assistance will be assembled to go over your project. They have various backgrounds—MEMS technicians and staff, electrical and MEMS design engineers, facilitators, imaging experts, and more, depending on the perceived needs of your project. This meeting is an idea brainstorming session, so expect things to be fast paced and disorganized at first as people throw out their thoughts and ideas. I found it fascinating to watch the opinions coalesce, the white board filling up with sketches, different people grabbing a dry erase marker to contribute to the concepts. After about an hour, we had a clear path forward for my research goals, had assigned me a staff mentor, and had arranged the first training sessions.

Meetings

You are going to need regular follow-up meetings as you come across new challenges and/or refine your design. I had a regular weekly meeting with my staff mentor, but I also scheduled discussions with various staff members with specific expertise—such as wireless communication and SEM usage.

Most labs have weekly or biweekly "drop-in" meetings as well where you can ask about anything on your mind, from design to processes to tool usage. Whether you have a specific project question or not, these group discussions are invaluable. You get to meet other lab users, learn about their projects, and learn who the subject matter experts are that you might want to connect with in the future. Bring a notebook! This is a great place to start building your list of resources.

Since everyone is incredibly busy, it is best to schedule appointments with anyone you want to meet with. Do this well in advance, going through their assistant if that is the preferred method. I have yet to meet a lab director that was in charge of their own calendar, so go to the source and get scheduled in.

Training

Most training is a two-part process. You can complete the first part online on your own time. It usually includes reading, videos, and quizzes. At the end, there is some kind of test you need to pass before you can proceed with the second part of the training, which is in the cleanroom on the equipment.

The in-person training requires an available staff member <u>and</u> an available piece of equipment, so you need to schedule this online considering both these criteria. Usually, your staff liaison will do the scheduling for you, either being the trainer themselves or recruiting the trainer on your behalf. They will also block out the time and reserve the equipment. You will want to confirm with the trainer the day before that everything is still proceeding according to plan, as equipment breaks down, processes exceed their time allotment, and people forget things and like reminders.

Your first in-person training is part demonstration and part you doing the work. You will be able to ask lots of questions and take notes in your cleanroom notebook. However, some time may pass before you are ready to work on your actual project, so the first time you use a piece of equipment on your own, you will have someone looking over your shoulder to assist, in case you have any questions or concerns about proper operation.

Time on Equipment

Once you are trained, your profile gets updated for that piece of equipment, and you can schedule time to use it. This is done online and should only be done when you are confident you will need the equipment at that time. Do not schedule multiple processes well in advance, as if you are like me, your optimism will not correlate well with reality—everything takes longer than anticipated. If you do schedule to use some equipment and find you will not need it (or are not ready yet), cancel as soon as possible so someone else can use it. Like a doctor's appointment, try to give 24 hours' notice.

When you sit down to use a piece of equipment, the first thing you have to do is login to start the time running. Some equipment is very expensive to use, so come prepared to get started immediately. When done, be sure to logout so you are not charged for time not used.

Be courteous of others and always be logged out and cleaned up in time for the next user to start on time.

Demonstrations

There may be some equipment that you want to see demonstrated. You can ask around and find out if anyone will be using that equipment soon and then ask if you can hover and ask questions, or you can ask a staff member to demonstrate its use, even if you have no intention of using it yourself or have yet to be trained. This is especially useful when considering different options in your design process, such as sputtering vs. CVD.

Another time to get demonstrations is when you feel unqualified or uncomfortable with the use of some equipment. Perhaps you are collaborating with someone and will not be operating the equipment yourself, but you still want to be familiar with its use. Everyone tends to be collaborative in the labs, so feel free to ask for the assistance you need.

Chapter 16
Beyond NNCI: International Facilities

For those outside the United States, there are some excellent open-use, national laboratories around the world. Compiled in this chapter are several laboratory options with details on how to contact them for further information. This list is far from comprehensive, but is a good start for those seeking to use international MEMS fabrication facilities.

Australia

Australian National Fabrication Facility (ANFF)

Website: http://www.anff.org.au/

Investigator(s):	Dr Jane Fitzpatrick, Acting CEO and Chief Operations Officer Tom Eddershaw, Marketing and Communications Matthew Wright, Office Manager
Sponsor:	ANFF Headquarters Registered office: 151 Wellington road, Clayton VIC 3168

Abstract

The ANFF was established under the National Collaborative Research Infrastructure Strategy (NCRIS) and consists of eight nodes and includes researchers and partners from 21 institutions. ANFF provides services for both academic researchers and industry. Researchers are able to either gain direct access to facilities under expert guidance, contract for specialized products to be made, or undertake contract research projects.

Each node provides their facilities on an open-access basis for researchers to engage in interdisciplinary research across the following fields:

- Micro- and nanoelectronics.
- Microfluidics and MEMS.
- Bio-nano applications.
- Advanced materials.
- Sensors and medical devices.
- Photonics.

Equipment and resource capabilities are listed at http://www.anff.org.au/capabilities.html.

The nodes are hosted at the following institutions:

1. *Victorian Node*—Monash University, Melbourne Centre for Nanofabrication. Partner institutions: University of Melbourne, Deakin University, LaTrobe University, Swinburne University, RMIT (Royal Melbourne Institute of Technology) University, and CSIRO (Commonwealth Scientific and Industrial Research Organisation) Agency.
2. *Australian Capital Territory (ACT) Node*—Australian National University.
3. *Western Australia Node*—University of Western Australia Crawley. Incorporates: Microelectronics Research Group (MRG).
4. *Queensland Node*—University of Queensland and Griffith University. Laboratories: Soft Materials Processing Facility at the Australian Institute for Bioengineering and Nanotechnology (AIBN), the BioNano Device Fabrication Facility at the Centre for Organic Photonics and Electronics (COPE), and Queensland Microtechnology Facility as well as the Raman Spectroscopy Facility at the Queensland Micro- and Nanotechnology Centre (QMNC).
5. *New South Wales Node*—University of New South Wales.
6. *South Australia Node*—University of South Australia. Partner institutions: Flinders University.
7. *OptoFab Node*—Macquarie University. Partner institutions: University of Adelaide, University of Sydney and Bandwidth Foundry International.
8. *Materials Node*—University of Newcastle. Partner institutions and laboratories: University of Wollongong's Australian Institute for Innovative Materials laboratory, Intelligent Polymer Research Institute (IPRI), and Institute for Superconducting and Electronic Materials (ISEM), as well as the University of Newcastle's Centre for Organic Electronics.

Collaborating

It is a stated goal of the ANFF to foster international collaboration. According to CEO Dr. Jane Fitzpatrick, they aim to "provide access to both tools and expertise on an open-access model to anyone who needs it." Researchers should contact the relevant node and submit a short proposal. If the interested party does not know which node would be most suitable, they are invited to contact the ANFF headquarters for guidance. Nodes aim to provide new users access within 1 month of application. Dr. Fitzpatrick further stated that they "have a number of modes of access but the preferred one in many places is that they train the users to the level they need to get the results they want with expert support and guidance along the way."

There is an initial contact form available on the website where the user identifies as a PhD student, publicly funded, or industry researcher. They then describe their project and submit it to the appropriate node.

Information about access and pricing is provided on the website as a PDF document. Some of the pricing information is excerpted below (Table 16.1):

Australia is also host to the biennial International Conference on Nanoscience and Nanotechnology (ICONN). This conference is held every other year in February in even number years (https://www.iconn2020.com/). ICONN features a diverse array of multidisciplinary talks designed to connect world-leading scientists, students, engineers, industry participants, and entrepreneurs working in the field of nanoscale science and technology to discuss new and exciting advances in the field.

Brazil

Brazilian Nanotechnology National Laboratory (LNNano)

Website: https://lnnano.cnpem.br

Investigator(s):	Adalberto Fazzio, Director
	Edson Leite, Scientific Director
Sponsor:	Brazilian Center for Research in Energy and Materials (CNPEM)Giuseppe Max Scolfaro Street
	10,000 - High Tech Polo II of Campinas
	Sao Paolo, Brazil
	ZIP Code 13083–970

Table 16.1 Sample prices for ANFF Victoria Node

Equipment at Victoria Node	Academic or publicly funded	Industry
Flagship equipment (i.e., electron beam lithography)	$90 AUD/hour ($750 cap per 24 hrs)	$225 AUD/hour ($1875 cap per 24 hrs)
Tier 1 equipment (i.e., atomic force microscope)	$70 AUD/hour	$170 AUD/hour
Tier 2 equipment (i.e., optical profilometer)	$45 AUD/hour	$115 AUD/hour

Abstract

The LNNano is one of four national laboratories under the Brazilian Center for Research in Energy and Materials (CNPEM). The other three are the National Synchrotron Light Laboratory, National Biosciences Laboratory, and National Biorenewable Laboratory. It is also the headquarters for the Binational Brazil-China Center of Nanotechnology with its partner, the Chinese Sciences Academy.

The LNNano was created in 2011 and incorporates several distinct laboratories with different capabilities. They provide multiple services to users, including R&D, consultancy, and customized training for individuals, groups, or companies:

- Electron microscopy (LME).
- Laboratory for surface science (LCS).
- Thin films and microfabrication (LMF).
- Metal characterization and processing laboratory (CPM).
- Nanostructured soft materials (LMN).
- Single-particle CryoEM.
- Functional devices and systems group.
- Nanobiotechnology and nanotoxicology group (NBT).
- Special functional and advanced dispositives (DCS).
- Support to business and industrial partners.

The last is actually a support group to facilitate use by industry partners and fits well with their mission statement regarding partnership on projects (translated):

LNNano is fully willing to associate to start-ups, nascent companies or consolidated companies of any size and business nature to act as a partner in short and long-term R&D projects.

Any LNNano partner in projects has the prospect to have its financial participation partially funded by development agencies to which LNNano is associated, which have specific programs or open calls for the funding of R&D projects in nanotechnology. Examples of these agencies are Embrapii, Sibratec, BNDES, FINEP, FAPESP among others.

Partners can also take advantage of funds from thematic networks specifically focused on nanotechnology, such as the Brazilian Nanotechnology System – SisNano, or from funds originated from cooperation programs between LNNano, and Brazilian and abroad R & D institutions.

Collaborating

To request services, such as to have LNNano develop your MEMS device, you must fill out a Request for Services form available on their website. https://lnnano.cnpem. br/innovation-initiatives/innovation-qa/.

LNNano wants to increase user engagement and offers tours on Wednesday to high school, college, and academics to foster interest in nanotechnology. Tours last 1.5 hours and are meant to accommodate groups of 15–20 people. They also offer scholarships, summer internships, workshops, and seminars as part of their comprehensive educational suite. https://lnnano.cnpem.br/education/educational-resources/.

Industry users must request LNNano to conduct research and fabricate MEMS devices on their behalf per the Request for Services form listed above.

All new academic users must register and submit a proposal through the User Portal at https://portal2.cnpem.br/cadastro/login.jsf. Before submitting a research proposal, users are advised to visit the relevant LNNano facility webpage to review the available services and technologies, as well as recommendations about sample preparations, protocols, and other detailed specifications about the research proposal submission process.

Once the research proposal is approved, the user will receive an email with instructions for the following three steps to be carried out prior to working at LNNano facilities:

1. Confirm the team and delegate responsibility (Confirme a equipe de campo).
2. Accept Commitment Term (Aceite o termo de Compromisso).
3. Safety Training (Responda o questionário de treinamento de segurança).

China

Ministry of Education (MOE) Key Laboratory of Thin Film and Microfabrication Technology

Website: http://en.sjtu.edu.cn/research/centers-labs/
moe-key-laboratory-of-thin-film-and-microfabrication-technology/

Investigator(s):	Professor Zhang Yafei, Director
	Professor Zhuang Songlin, Director of Academic Committee
Sponsor:	Shanghai Jiao Tong University
	International Science and Technology Project Office
	800 Dongchuan Rd. Minhang District
	Shanghai, China

Abstract

The MOE Key Laboratory's mission is to work on basic and applied research for cutting-edge micro- and nanoscience and technology, by engaging in thin-film electronic materials and thin-film sensors, microfabrication technology, microelectromechanical systems (MEMS), nanoelectronics, and nanomaterials and technology. Its key areas of research are as follows:

- Non-silicon microfabrication and MEMS.
- Nanofabrication and device design.
- Nano-bio and medical technology.

Collaborating

As part of the Chinese Sciences Academy, Shanghai Jiao Tong University has a long history of collaboration with prestigious international and US universities, including MIT and UC Berkeley. They have launched dozens of companies and have undertaken over 900 projects over the years, resulting in several dozen patents.

It is unclear from the website how one goes about establishing a research collaboration to work with the MOE Key Laboratory, so it is suggested that interested parties contact the Lab Director, Dr. Yafei, for more information.

Denmark

National Centre for Nano Fabrication and Characterization (DTU Nanolab)

Website: https://www.nanolab.dtu.dk/english/aboutus

Investigator(s):	Jorg Hubner, Director
	Anders Jorgensen, Deputy Director and Head of Customer Support
	Leif Johansen, Head of Operations
	Flemming Jensen, Head Of Process Engineering
	Jakob Birkedal Wagner, Scientific Director
Sponsor:	Technical University of Denmark (DTU)
	National Centre for Nano Fabrication and Characterization
	Oersteds Plads – Building 347
	DK-2800 Kongens Lyngby
	Denmark

Abstract

DTU Nanolab is an open use facility with over 500 users from academia and industry, including international users. Their laboratory expertise is research-based, including viewing chemical reactions with atomic resolution. More than 30 companies use their open-access facilities and expertise to develop devices and conduct small-scale production run testing.

Their cleanroom facility features the following activities and prides itself on flexibility and cooperation:

- Direct access to DTU Nanolab's wide range of equipment and expertise.
- Installation of customer-owned equipment.
- Rent of restricted access areas inside the cleanroom.
- Insourcing of development and production to carry out complex development and production work.
- Quality assurance and controlled fabrication environment.
- Consultation services in connection with process design, product development, and production.
- Education of bachelor, master, and PhD students as well as education and training for industrial customers.

DTU Nanolab's Microscope Facility is impressive with a suite of seven electron miscroscopes—four scanning electron microscopes (SEM), two of which are dual beam, and three transmission electron microscopes (TEM).

Pricing for microscope and cleanroom equipment use is available in the price-book, which is a PDF that can be accessed here, https://www.nanolab.dtu.dk/english/Microscopes/FAQ/Prices. A table from the pricebook is excerpted below for reference. All prices are in Danish kroner (10 kroner equals about 1 US dollar) (Table 16.2).

Research

DTU Nanolab has several special interest research groups:

- Molecular windows.
- Biomaterials microsystems.
- Polymer micro- and nanoengineering.
- Silicon microtechnology.
- Microscopy.

Users with interest in these research focuses should click on the relevant link under the Research tab. Access to the "Silicon Microtechnology" group is via http://nanolab.dtu.dk/english/Research/Silicon-Microtechnology, where you can find information about collaborations, publications, current research, and more.

Table 16.2 Sample prices for DTU Nanolab

Service from Nanolab	Unit	Commercial activity	External project work, Danish academia	DTU Partner with budget in external projects	Internal DTU projects
Cleanroom access (below cap)	Kr/hour	800	255 + 44% OH	255	0
Category A tools	Kr/hour	410	125 + 44% OH	125	0
Category B tools	Kr/hour	650	330 + 44% OH	330	0
Category C tools	Kr/hour	3600	840 + 44% OH	840	0
Category D tools	Kr/hour	1200	240 + 44% OH	240	0
Category E tools	Kr/hour	1700	415 + 44% OH	415	0
Category P tools	Kr/hour	410	0	0	0
Nanolab assistance	Kr/hour	1350	330 + 44% OH	330	0
Area rent	Kr/m²/ imonth	2000	NA	NA	NA
Materials		At cost+20%	At cost + 44% OH	At cost	At cost

Collaborating

DTU Nanolab welcomes all types of users from academia and industry. The lab will engage with users in any number of ways, including research and development, prototyping, consulting, small-scale production, and education. They invite guest researchers to the lab, and DTU offers international unpaid guest researchers and PhD students assistance with administrative, practical, and cultural issues before and after arrival. They will even provide assistance with the process for a work or residence permit and/or EU residence documents.

France

Micro Nano Bio Technologies (MNBT)

Website: https://www.laas.fr/public/en/micro-nano-bio-technologies

Investigator(s):	Bernard Legrand, Head Anais Moulis, Secretary
Sponsor:	Laboratory for Analysis and Architecture of Systems (LAAS) 7, avenue du Colonel Roche BP 54200 31,031 Toulouse cedex 4, France

Abstract

The MNBT laboratory is part of the national Laboratory for Analysis and Architecture of Systems (LAAS) in Toulouse, France. It is leading research activities at the intersection of materials engineering, applied physics, and life sciences with an objective to deliver enabling micro- and nanotechnologies in biological and environmental applications. Some of their current research foci are as follows:

- Materials science investigation of electronic materials.
- Device design and fabrication.
- Multiplexed biofunctionalization techniques of surfaces at the nanoscale.
- Probing living matter in real time with micro- and nanoscale resolution (e.g., sensing based on electrochemistry, electromagnetism, and optics).

Their research is divided into several groups, many of which have links to their own webpages:

- ELiA (engineering in life sciences applications).
- MEMS (microelectromechanical systems).
- MH2F (fluidic high-frequency micro and nano).
- MICA (microsystems analysis).
- MPN (materials and processes for nanoelectronics).
- NBS (nanobiosystems).

One such page, MEMS, at https://www.laas.fr/public/en/MEMS gives further information about research topics, team members, recent publications, PhD dissertations, current and completed research contracts, and opportunities to join the MEMS group as a PhD student or intern.

Collaborating

LAAS features many interesting ways to collaborate, including the Affiliates Club, which is a service providing technical and scientific information to the members of the club. There is also a webpage devoted to start-ups launched by LAAS-CNRS that provides links to several recent start-up success stories, https://www.laas.fr/public/en/start-ups. A stated goal is to support "its researchers, engineers, technicians and PhD students in creating their own business and assists them during the first years of their existence by providing technological and local support, and hosting them during the incubation phase."

A list of all equipment in the MNBT lab is available via an online portal, which is accessible to registered users https://lims.laas.fr/WebForms/Equipment/EquipmentList.aspx.

To collaborate with the Micro and Nano Technology lab, information is provided in a guidebook downloadable as a PDF from https://www.laas.fr/public/en/micro-and-nanotechnologies-platform. It includes information about Renatech, which is a network of labs within France. Information from the guidebook is excerpted here:

- Hyperlinks are provided to apply to use the labs and register a project.
- Additional hyperlinks link to safety training and protocols.
- A training plan and calendar are provided to get scheduled after you obtain approval.
- Information on booking equipment, ordering supplies, and touring the clean-rooms is also provided.

Netherlands

MESA+ Institute Nanolab

Website: https://www.utwente.nl/en/mesaplus/

Investigator(s):	Albert van den Berg, Scientific Director
	Guus Rijnders, Scientific Director
	Janneke Hoedemaekers, Managing Director
Sponsor:	University of Twente
	MESA+ Institute Nanolab Building
	Hallenweg 15, 7522 NH Enschede
	The Netherlands

Abstract

The MESA+ Institute is a leading nanotechnology research institute with a focus on key enabling technologies (KETs) in the following areas:

- Photonics.
- Fluidics.
- Hard and soft materials.
- Devices.

Their main areas of research focus are in health, ICT, and sustainability. For health, their areas of research are in diagnostics and biomarkers of disease. In ICT, their research involves developing energy-efficient computational power needed for highly challenging applications. In sustainability, they focus on energy storage applications.

Collaborating

The MESA+ Institute actively seeks cooperative alliances with business and industry, (semi)government, nongovernment organizations (NGOs), and knowledge institutes. Contact information to discuss your project https://www.utwente.nl/en/mesaplus/about/organization-governance/.

MESA+ specializes in the following:

- Matching experts for your project: They will put you in touch with the right contact person and/or researcher at the MESA+ Institute.
- Business development: They will help you to create impact with the results of your research by helping bring them to the market.
- Funding acquisition: They will provide strategic advice, possible funding opportunities, planning, and funding proposal development for your project.

They have a tech transfer office to help you convert your idea into a spinoff or licensed technology. For details on how to discuss these options and support services, contact them from https://www.utwente.nl/en/mesaplus/innovation/tech-transfer/.

New Zealand

University of Canterbury Nanostructure Engineering, Science and Technology Research Group

Website: https://www.canterbury.ac.nz/research/facilities-and-equipment/research-equipment-and-facilities/electrical-and-computer-engineering-labs/

Investigator(s):	Maan Alkaisi, Director Martin Allen, Co-Director Volker Nock, Co-Director
Sponsor:	University of Canterbury College of Engineering Electrical and Computer Engineering Level 5 and 3 Link Building, Private Bag 4800 Christchurch 8020 New Zealand

Abstract

The Electrical and Computer Engineering's (ECE) Nanofabrication Laboratory contains facilities for semiconductor material processing, nanofabrication, sensor, and microfluidic device development. This is the national micro- and nanofabrication facility for New Zealand and is the main fabrication facility for the interdepartmental Nanostructure Engineering, Science and Technology (NEST) research group (https://www.canterbury.ac.nz/engineering/schools/ece/research/nest/), as well as the MacDiarmid Institute for Advanced Materials and Nanotechnology (https://www.macdiarmid.ac.nz/).

Research interests are focused on the following:

- Functional nanostructures.
- Materials for energy capture and utilization.
- Tomorrow's electronic devices.

Collaborating

The ECE Nanofabrication Lab is a small facility with a variety of micro- and nanofabrication capabilities. It is most suitable for early stage investigations and small production runs. Users are expected to conduct their own fabrication; however, there is training and support for purchase of supplies. Pricing for academics is on a per annum user fee. Industrial users not affiliated with the MacDiarmid Institute should contact the lab director for more information on how to gain access to the facility.

Portugal/Spain

International Iberian Nanotechnology Laboratory (INL)

Website: https://inl.int/

Investigator(s):	Lars Montelius, Director-General
	Paulo Freitas, Deputy-Director General
Sponsor:	Governments of Portugal and Spain
	INL – International Iberian Nanotechnology Laboratory
	Avenida Mestre José Veiga s/n, 4715–330
	Braga, Portugal

Abstract

The INL is a joint laboratory between Portugal and Spain that serves as an international hub for the development of nanotechnology. It is divided into six departments:

- Nanoelectronic engineering.
- Life sciences.
- Quantum and energy materials.
- Micro- and nanofabrication.
- Nanophotonics.
- Advanced electron microscopy and spectroscopy.

Their applied research areas are comprehensive, with broad coverage of the following:

- Health—INL is focused on the development of novel technologies for the early diagnosis and treatment of diseases.

  - Diagnostics.
  - Therapeutics.

- Food and environment—INL covers complementary fields ranging from food technology to biology and chemistry, enabling a versatile approach to address challenges in the food quality and safety and in environmental monitoring.

  - Lab-on-chip.
  - Smart packaging.
  - Nanomaterials.

- Information and communication technologies—INL activity is focused on 5 technologies: spintronics, graphene, thin films, MEMS, and CMOS IC design.
- Renewable energy—INL performs research and development along a wide range of research lines, including solar energy harvesting, conversion, and storage.

INL has specific interest in developing commercial enterprises and states, "We facilitate the development and uptake of new ideas and transfer them into commercial value that will increase improve industry's productivity, competitiveness and foster economic growth." To foster this, they provide customized services for each project and the highest level of confidentiality. They also offer an international market square for nanotechnology and key enabling technologies (KETs). Through INL, partners can easily reach international governments and countries. They also have an incubator program and space for up to ten start-up companies to be located on site.

Collaborating

To obtain access, users must first fill out an application form available at https://inl. int/user-facilities/user-access/. After approval, user will receive an email with login credentials to access the Laboratory Information Management System http://lims. inl.int. This platform supports the management of all equipment, including training request, booking, logging, and billing.

South Africa

National Centre for Nano-Structured Materials (NCNSM)

Website: https://www.csirnano.co.za/

Investigator(s):	Suprakas Sinha Ray, Director and Chief Researcher
	Gugu Mhlongo, Facility Manager
Sponsor:	CSIR Building 19B Scientia Campus
	Meiring Naude Road
	Brummeria, Pretoria 0184
	South Africa

Abstract

The NCNSM Characterization Facility run by the Council for Scientific and Industry Research (CSIR) provides a wide range of instrumentation for use by the nanotechnology community, and others, to characterize their research samples. They welcome other universities, researchers, and industry to use their facilities.
    Active research within the facility includes the following:

- Advanced Materials for Device Applications—This group focuses on the development of nanomaterials that can ultimately be used in an array of sensors such as gas sensors.
- Nanomaterials Industrial Development Program—This program focuses on the development of new and advanced materials through the incorporation of nanotechnology.
- Nanomaterials for Water Remediation Project.
- Catalysis research—This project is developing biomass-derived green chemicals, fuel blending chemicals, and biofuel using nanocatalytic processes.

The NCNSM has numerous local collaborators, including the following:

- University of Cape Town.
- University of the Free State.
- University of Witwatersrand.
- University of Western Cape.
- iThemba Laboratories,
- University of Pretoria.

In addition, they have at least ten international collaborators from around the world and three featured industry collaborators.

The NCNSM links to several other micro- and nanotechnology initiatives within South Africa https://www.csirnano.co.za/about-us/links/, including the following:

- Main CSIR website.
- The Department of Science and Technology.
- The National Nanotechnology Strategy.
- South African Nanotechnology Initiative.
- Microscopy Society of Southern Africa.

Collaborating

To initiate contact with NCNSM, use the contact form, https://www.csirnano.co.za/about-us/contact-us/, and they will reply promptly. In addition to the contact form, this link also provides email addresses and phone numbers for key individuals.

Sweden

The Electrum and Albanova Laboratories at KTH University

Website: https://www.kth.se/en/eecs/forskning/electrumlaboratoriet-1.262770

Investigator(s):	Göran Stemme, Head of Devision
	Wouter Metsola van der Wijngaart, Deputy Head of Division
Sponsor:	Division of Micro and Nanosystems
	KTH School of Electrical Engineering and Computer Science
	Malvinas väg 10
	SE-100 44 Stockholm
	Sweden

Abstract

The Electrum Laboratory and Albanova Nanofabrication Facility, housed within Electrical Engineering and Computer Science at KTH University, offers complete processes for nano- and micro-manufacturing. The laboratories are operated by KTH and Acreo, and both support projects from start-up to mature technologies.

All resources are available as open access, including the following:

- A fully equipped cleanroom for device research and manufacturing.
- A flexible cleanroom environment for materials and device-oriented research and development.
- World-class characterization laboratories.
- Sophisticated software for calculations, simulation, and design.

Their staff is highly skilled and available for users' projects:

- Process and development services.
- Commissioned research and development.
- Prototyping and small-scale production.
- Courses in process technology, characterization, and cleanroom infrastructure.

The major research field of the division is micro- and nanoelectromechanical systems (MEMS/NEMS), where micro- and nanofabrication techniques and materials are adapted to the making of small, low-cost, high-performance electromechanical, optoelectromechanical, RF/microwave, and micro- and nanofluidic devices. Research areas of interest in micro- and nanotechnology include the following:

- Biomedical microtechnology.
- Micro and nanofluidics.
- RF microwave and terahertz microsystems.
- NEMS and nanosystems.
- Organ-on-a-chip and cell models.
- Photonics.
- Sensors.
- Soft materials.

Collaborating

KTH's Micro and Nano Laboratories actively encourages outsider researchers and companies to join their lab, and they state, "Collaborating with KTH gives you an excellent opportunity to develop the knowledge base of your company, gain new perspectives on research questions and access an advanced infrastructure for research and innovation in an international environment. We regularly collaborate with external corporations and organisations and base our questions on the needs of relevant industries."

Users have the option of collaborating with students, alumni, research personnel, and research groups or centers. https://www.kth.se/en/samverkan/samarbeta-med-oss/samverka-med-forskar.

The AlbaNova Lab is where users can train to use a variety of micro- and nano-fabrication equipment. Usage, training, and booking are all managed through the Lab Information Management System (LIMS), which can be downloaded as a PDF at http://www.nanophys.kth.se/nanophys/facilities/nfl/nfl-frames.html.

The academic user flat rate per year is 55,000 SEK, which converts to about $5500 USD per year. Ten Sweden krona equal about one US dollar. Occasional or infrequent users can pay per use per myFAB rate structure. Without registering for the lab, it is not possible to determine the fees for industry use.

United Arab Emirates

Khalifa Semiconductor Research Center (KSRC)

Website: https://www.ku.ac.ae/khalifa-semiconductor-research-center-ksrc/

Investigator(s):	Mohammed Ismail, Chair
Sponsor:	Khalifa University
	P.O. Box: 127788
	Abu Dhabi, UAE
	https://www.ku.ac.ae/campus/location-map/

Abstract

Khalifa University's Semiconductor Research Center has three branches: nanotechnology, biomedical applications, and system-on-chip. Their focus on nanotechnology addresses growing interest in scaling down basic electronic devices to a nano range, as well as applications for power saving, small integrated devices, and high speed. This corresponds with needs identified by the International Technology Roadmap for Semiconductors (ITRS) and in the Abu Dhabi 2030 strategic plan.

KSRC researchers have published a book titled "Energy Harvesting for Self-Powered Wearable Devices." The book is a reference for design engineers, practitioners, scientists, and marketing managers in the semiconductor industry developing integrated, self-powered, platform system-on-chip solutions for wearable devices. https://www.springer.com/gp/book/9783319625775

KSRC nanotechnology-related research projects include the following:

- Energy loss minimization for magnetic tunnel junctions.
- Minimization of contact energy losses with end and side contacts for nanowires.
- All-optical quantum random generator driven by spin noise in semiconductor nanostructures.
- Low-power devices based on nano Schottky junctions.
- Energy efficient nano-contacts and for spintronics.
- Functional devices based on graphene composites.
- Development of a nanoparticle-based transistor.
- Wide band-gap semiconductors for high-power, high-temperature, and high-speed applications.

Collaborating

Collaboration occurs at the university level. Khalifa University is actively involved in establishing long-term strategic partnerships with a variety of organizations. The university currently enjoys strong relationships with over 20 national and international partners including industry leaders, multinational entities, government agencies, universities, and other institutions. They use these partnerships to facilitate solving applied problems.

In response to Abu Dhabi's 2030 strategic goals, Kahlifa University has undertaken several broad university research initiatives and partnerships focused on the following areas: aerospace, biomedical, healthcare, national security, telecommunications, information technology, and nuclear energy.

They have set up a business development unit dedicated to working with outside collaborators, https://www.ku.ac.ae/business/.

They offer both consultancy services, https://www.ku.ac.ae/consultancy-services/, and laboratory rental https://www.ku.ac.ae/laboratory-rental/. To obtain more information about how an external user could gain access to the micro- and nanofabrication facilities, contact them via email at BusinessDevelopment@ku.ac.ae.

United Kingdom (Scotland)

James Watt Nanofabrication Centre

Website: http://www.jwnc.gla.ac.uk/index.html

Investigator(s):	Iain Thayne, Director
	Arthur Smith, Operations Manager
	Brendan Casey, Commercial Enquiries (Kelvin Nanotechnology Ltd)

Sponsor:	University of Glasgow
	School of Engineering
	Rankine Building
	Oakfield Avenue
	Glasgow, G12 8LT
	U.K.

Abstract

The Watt Nanofabrication Centre is a long-standing and well-known facility with over 35 years of experience delivering micro- and nanofabrication solutions. One of their specialties is in electron beam lithography to develop solutions in processing, nanotechnology, nanoelectronics, optoelectronics, bioengineering, biotechnology, lab-on-a-chip, clean tech, energy, and photovoltaics.

The James Watt Centre undertakes fundamental, applied, and commercial research and can develop devices for small production runs.

The James Watt Centre collaborates with over 90 different international universities and research institutes along with working with over 288 companies from 28 countries around the world.

Details on their equipment can be found at http://www.jwnc.gla.ac.uk/equipment.html and is categorized by the following:

- Electron beam lithography.
- Nanoimprint lithography.
- Nanoinjection molding.
- Optical lithography.
- Metal deposition.
- Dry etch.
- PECVD insulators.
- Metrology.
- Electron microscopes.
- Thermal processing.
- Miscellaneous.
- Electrical.
- Optical.
- Biological.

Collaborating

A large number of academic institutions already partner with the James Watt Nanofabrication Centre, and the center also conducts collaborative research with many government organizations, such as EPSRC, BBSRC, TSB, EC, DARPA, SRC, and Wellcome Trust.

Industry users access the center through Kelvin Nanotechnology (KNT) https:// www.kntnano.com/. Services range from access to fabrication to consultancy. It is also possible to consult with an academic at Glasgow University, http://www.jwnc. gla.ac.uk/consultancy.html.

To gain access to the lab, users should contact either the James Watt Nanofabrication Centre director or the director for Kelvin Nanotechnology at http:// www.jwnc.gla.ac.uk/contact.html.

Chapter 17
Intellectual Property

Ownership of IP

Although housed at universities, the NNCI facilities are open use, and the universities will not make any claims of ownership on your intellectual property. Essentially, you are renting time in their facilities. As with anything, do not disclose what you do not want people to know. For those aspects you need to disclose, you need to have a nondisclosure agreement in place with the facility (if allowed).

If new IP for your project is developed while using a facility, even if with the help of staff, it belongs to you. In the extremely rare instance that a new fabrication technology is developed by staff for use on your project, then that fabrication technology would be available to other users. However, your use of that technology would not be disclosed.

Nondisclosure Agreements

Below is the NDA used at WNF, taken from the user manual. Other facilities, like CNS at Harvard University, do not do NDAs due to the amount of people using their facility, so you will have to ask what the policies are at your closest facility.

Intellectual Property and Security

While working in the WNF, you will not have intellectual property (IP) restrictions or entanglement with the University of Washington. Many clients execute an NDA in order to protect their IP (use of the UW preapproved form will expedite the NDA process).

© Springer Nature Switzerland AG 2019
D. Munro, *DIY MEMS*, https://doi.org/10.1007/978-3-030-33073-6_17

Occasionally staff will engage in collaborative development campaigns with users. In these cases, general processing techniques that are not IP-specific may be shared with the general user base, but applications and full process flows will not be shared unless given explicit permission.

Collaboration Versus Protecting IP

I am rather fond of the policy at CNS, which is based on collaboration. Although the risk theoretically exists that someone will steal your idea and develop it for themselves, in practice, that has not happened. Everyone's interests are very unique to their specific application, but the MEMS processes are shared. We are all trying to use this new technology for microfabrication, and we are all somewhere along the path of learning how to do it the best we can. The NNCI labs are great places to learn the latest techniques and personal successes. My philosophy is, "If they can do it, I can do it," and rather than start from scratch like them, I can wholesale borrow their process and adapt it to my application.

However, if this is a concern for your company, choose a facility that does NDAs, keep your IP and project ideas close to your chest, and only disclose things on a need to know basis.

Named Inventors

For a patent, you will need to disclose your device design, and that may include some of the MEMS processes you have developed. In spite of this, staff that have helped you with those processes are <u>not</u> named inventors. These facilities are government-sponsored labs, so they do not claim any ownership on your IP. Although often housed at universities, the university has no ownership of the facilities, and they will not claim any ownership of your IP, either.

The exception is if you have a collaborator, either at the lab or at the university. When establishing an agreement with a collaborator, spell out clearly (in advance) who owns the IP that is developed.

Patenting

Patents are a necessary evil, in many cases. You have to protect your ideas from your competitors in order to gain a marketing advantage, and they are used to value the worth of your company. But over 90% of patents are never developed into a product, and they are expensive, time consuming, and fraught with litigation if you patent anything worthwhile.

With that said, I will admit it is very cool to have several patents myself, mostly around the research I have done with MEMS sensors. Having those patents has opened a lot of doors and allowed me to pursue research that I am passionate about. I look forward to learning what great new products you develop using MEMS technology, too!

Chapter 18
Getting Project Assistance

On-Site Incubator Spaces

Each facility typically has a small set of experienced users who are available as independent contractors on a short- or long-term basis. These "contractors" do not work for the facility, but are available for private hire. This can shorten the learning curve and provide additional intellectual property isolation within the facility.

You may also do a sponsored research program. Companies may establish a sponsored research program with any faculty member to gain access to that faculty member's expertise. This is typically a longer-term collaboration that does not involve specific deliverables. Agreements on intellectual property are between the company and the faculty member. The faculty member will also have experienced students who can be hired to assist with your project at rates that are often more cost-effective than using independent contractors. The facility has essentially no part in arrangements with faculty members.

Hourly Technical Assistance

You will need technical assistance, and there is no better help than the lab staff—all of them experts in their subject matter. No one's time is free, but the rate for help is surprisingly affordable. Although they can usually only devote a few hours to your project, that can be an invaluable investment. For my project, I used DRIE, which is difficult to dial in. My staff liaison, Dr. Andrew Lingley, tried multiple iterations on my behalf, altering the film coatings and settings until we were successful. Because each DRIE process took hours to complete, he would run them after hours and then come pull them out before others needed the equipment. It was an incredibly worthwhile investment that I very much appreciated.

© Springer Nature Switzerland AG 2019
D. Munro, *DIY MEMS*, https://doi.org/10.1007/978-3-030-33073-6_18

Working with an Academic Partner

If you cannot devote the time required to operate the equipment yourself and/or need a large amount of expert assistance, an academic partner might be the best option for you. Look for an academic researcher who is conducting research similar to your project, and then contact them about collaborating with you. If you do not know someone who is in your research area, then go through the NNCI lab manager, and they will be able to recommend a few potential collaboration partners at their university.

Whenever you work with an academic partner, your project will be governed by the rules and costs associated with the university. For many universities, the overhead rate is extremely high (upwards of 50%), but the stipend for the academic researcher is often very low. They will often delegate the day-to-day work on your project to a graduate student, so the money you pay will be supporting a stipend and tuition for a student (in case that makes you feel better about the overhead rate). The good news is that you will be paying academic rates for use of the lab, rather than the higher industry rate, so in the end, the cost of collaborating is not as expensive as it sounds at first.

Hiring a Student Directly

You can also hire a student at the university directly. Many labs have REU programs as well as undergraduate students they hire to help work in the labs. You can sponsor a student to work on your project at very affordable rates, but know that your project will proceed more slowly and will have more of a learning curve than if you hired an experienced person.

I hired engineering undergraduates to work at the lab for me, and they were great. They quickly learn how to use the equipment, and they can help with a lot of bulk tasks, like preparing wafers for your next process (doing all the washing, spin-drying, spin coating, and photolithography steps so that you are ready to go right away). They are also great at dicing, soldering, mounting, wire bonding, collecting measurements on a microscope, performing electrical tests, and taking images.

Graduate students are a better choice for design and processing steps, as they have chosen to conduct research in MEMS already and have had significantly more coursework and personal experience. If you are looking for a MEMS research partner who will follow your lead but also make their own contributions to your design and process, graduate students are a great option. The university has requirements on how much they must be paid, but it is much less than an academic researcher.

Bringing in Your Own Personnel

You are welcome to bring in your own employees or summer students to work in the lab. If you want to build up in-house expertise and/or keep IP confidential, your own employee may be the way to go. They can take all of the training and work on-site as much as needed to take your project all the way through to completion.

Handing Off the Project to Lab Personnel

It is also possible to hand off your project to lab personnel to be fabricated on your behalf, but this is the least desirable alternative. Lab staff are extremely busy, and your project really will not receive the personalized attention it deserves. Only *you* care about your project as the number one priority. To everyone at the lab, it is just one of many interesting projects, and it is human nature to focus on the project where the person who cares about it is holding it in your face, tears in their eyes (figuratively, of course).

Hiring Subcontractors

Another good option is working with a subcontractor that already has an established relationship with the lab. There are several small companies that do MEMS design and process development and then work with MEMS labs that may or may not be affiliated with an NNCI facility. These contractors are more expensive, but they have highly qualified personnel who will get your project done in a timely manner and with exceptional quality.

If you do not plan to become a MEMS fabricator, or you simply do not have the time or in-house resources to take on developing the MEMS device yourself, then hiring a subcontractor might be the best option for you. During your design review and planning, talk to the lab management and get recommendations on whom they have interfaced with before. You can also find out which vendors are currently working in the lab and start up an informal conversation with someone to get more information.

Chapter 19
Costs

The NNCI facilities are set up to offer affordable access to everyone. They offer leading-edge tools, training, and expertise in all disciplines of nanoscale science, engineering, and technology to academic, government, and industrial researchers and hands-on education and outreach events for novice users.

According to Dr. Khbeis, most facilities operate in about the same manner and have about 90% of the same capabilities. The Washington Nanofabrication Facility has more industry users than most, but other facilities are interested in expanding their industry clientele and have programs to encourage this.

For example, Cornell University is open 24 hours a day and provides "an interactive learning and practicing environment critical to successful cutting-edge research." External users typically spend a week or two at their facility to complete their work with strong staff support, but many projects can also be accomplished remotely.

Cornell has a program they call CNF Foundry Partners which is designed to help launch new companies and create high-tech jobs. They have cost reductions available to help companies establish new fabrication capabilities as well. So, take a look around and see what NNCI facility and program might be the best fit for you.

Rates

Rates depend on the facility and what category of user you are. Generally, industry users pay a higher rate than academics. Below are the published rates for WNF.

Non-UW Academic, Nonprofit Organizations, and Government Researchers employed by or with appointments at non-UW academic institutions, recognized nonprofit organizations, and state/federal government agencies or national labs. These candidates are eligible for the Academic rate plus an overhead charge of 15.6%.

© Springer Nature Switzerland AG 2019
D. Munro, *DIY MEMS*, https://doi.org/10.1007/978-3-030-33073-6_19

Industrial Any employees of nonacademic and nongovernment organizations who intend on using the WNF facility for self-directed work. These candidates are eligible to select between the Industrial Fixed or Industrial Full plus an overhead charge of 15.6%.

Remote Entities that solely seek to conduct limited scope process work that is to be conducted by WNF staff provided on a best-effort, time, and materials cost reimbursement basis that includes the prorated hourly equipment plus engineering rate ($150/HR) and overhead of 15.6%. Due to the experimental nature of work, WNF cannot guarantee quality assurance in meeting all remote work specifications, but will use best-effort methods to attempt to obtain a satisfactory result for the client.

Rate Plans

These rate plans are available to users as dictated in the user classification section. Each project/user combination must be enrolled in a rate plan. Users with multiple affiliations and projects may be in multiple differing rate plans, but all plan charges, caps, and restrictions will be addressed individually for each organization/project/ user combination. Stockroom supplies and precious metals are charged at cost plus nominal overhead for all plans.

Academic Full The academic rate plan is directed to provide cost-effective access for academic researchers and is subsidized by the UW and grant activity. The plan provides a usage cap with a sufficiently high monthly notch to enable cost prediction for grant proposals. All equipment use that exceeds the *monthly* notch limit is charged at the Post Notch Cap rate. The cap applies only to equipment use and access fees. Training and engineering assistance is not capped. Furthermore, to foster the development of process-centric research groups, principal investigators with multiple *active* (incurring more than $500/month in charges) Academic students will receive a monthly rate discount on standard equipment and access rates (2 students 10%, 3 students 15%, 4 students 20%, 5+ students 25%).

Industrial Full The full industrial rate plan is a standard industrial rate, uncapped plan for light or intermittent industrial users. All access and equipment use charges are billed at the industrial hourly rate plus overhead of 15.6%.

Industrial Fixed The fixed industrial rate plan is directed toward consistently high-volume users by providing a consistent monthly charge (for below notch utilization) with a 6-month use commitment with open enrollment in January and July or upon inception as a new user. The fixed rate includes monthly card fee, unlimited daily access fees, and all equipment charges below the *monthly* notch limit.

All equipment use that exceeds the notch limit is charged at the Post Notch Cap rate (33% of hourly rate) plus overhead. Organizations can have a mixed group of user slots between Industrial Fixed and Industrial Full plans, but must assign a specific project/user to the slot during open enrollment. Organizations with multiple fixed users can share notch limits among the *fixed* users, so the monthly usage before being assessed at the Post Notch Cap Rate for 2 fixed users is $15,000 and 3 fixed users is $22,500. Organizations with multiple fixed rate accounts will receive a monthly rate discount on the fixed rate fee (3–4 fixed users 10%, 5+ fixed users 15%).

Training Training will incur engineering staff charge ($150/HR) plus the equipment rate.

Staff Support Staff assistance for nonequipment issues will be charged at the engineering rate of $150/HR. Requests for staff support should be made via email to staff members.

Electron Beam Lithography (EBL) The UW JEOL JBX-6300 EBL is one of the most advanced and cost-effective direct write tools in the nation. Despite this, WNF incentivizes its utilization by introducing the cap and notch cap constructs to this tool for each separate project/user combination. The cap is at 60 hours per year with the notch at 80 hours. Remote EBL work is not eligible for cap.

Equipment Rates are from $30 to $210/HR for Industrial users and $10 to $70/HR for Academic users per Table 19.1.

Table 19.1 Rate plan and fee structure at WNF

Organization:	UW	Non-UW	Non-UW	Non-UW
Tier	Academic	Academic	Industrial	Industrial
Level	Full	Full	Full	Fixed
Overhead rate	0.0%	15.6%	15.6%	15.6%
Access fees:				
Month	$100	$100	$100	$5000
Daily	$25	$25	$75	$0
Caps:				
Cap/notch reset	Monthly	Monthly	Monthly	Monthly
Cap/month	$2000	$2000	N/A	N/A
Post notch cap rate	33%	33%	N/A	33%
Notch point ($)	$4200	$4200	N/A	$7500
Annual budgetary estimate (<notch)	$30,000	$35,000	N/A	$70,000

Equipment Time

The rates for usage of equipment vary from $30 to $210 per hour at WNF for industrial users, which is based upon the cost of purchasing that piece of equipment as well as the cost to operate and maintain it. You are only charged for the amount of time used (in 15 minutes' increments). The NNCI facilities are not set up to make a profit, and in fact, they operate at a loss, which is why the NSF sponsorship makes them viable. You going in to do MEMS fabrication helps justify this continued expenditure, so make use of this great resource that is available to you.

For specific equipment rates, go to the facility's website, where you can obtain detailed information.

Supplies and Materials

MEMS fabrication supplies and materials are surprisingly cheap. Silicon wafers range from a few dollars to perhaps tens of dollars each, and you can fabricate hundreds of devices on a single wafer. Other supplies, such as resistors, capacitors, op-amps, wire, and solder pads, are so cheap, and you are required to buy a few dozen just to reach a few dollars of expense. Batteries can be a few to several dollars each, but the expense of supplies is generally negligible.

You will have an initial investment in tools and personal equipment that will cost from $100 to $300, depending on what you need. Most of the tools and plastic storage boxes are available for purchase at the lab, but you can find them for purchase online, too.

If you outsource a process, such as making pattern masks, you will pay perhaps a few hundred dollars at most.

Overall, you may be pleasantly surprised at how affordable it is to do MEMS fabrication.

Chapter 20
Taking the First Steps

Where to Start

This is an exciting time in medical devices and orthopedics as new technologies become accessible for us to use in our products. MEMS is one tool that can bring increased functionality and capability to our designs, thus providing more diagnostic information and hopefully improving patient outcomes.

If you are ready to take the first step, use the information provided in this book to read about a nearby NNCI facility, and then contact them and set up a time to come visit and have a tour. Then, fill out the paperwork and set up a design review for your project and see what is possible.

Getting Initial Advice

It is a simple truth that "we don't know what we don't know." If you have a project that you think might benefit from MEMS technology, contacting an NNCI facility and setting up an initial brainstorming session with them to learn what might be possible is an excellent way to get started. They can fill your head with possibilities and ideas, and you can use that to refine your project idea into something you want to try fabricating.

Further Assistance

If after reading this book, you still have questions or want further assistance, feel free to reach out to me. Nothing is more flattering to an author than having a reader be inspired enough to ask for more! I will help you to the best of my ability, and I can guide you on where to find even more excellent information so that you are successful.

You absolutely can do it yourself with microelectromechanical systems, and I am excited to hear about your successes.

© Springer Nature Switzerland AG 2019
D. Munro, *DIY MEMS*, https://doi.org/10.1007/978-3-030-33073-6_20

Index

© Springer Nature Switzerland AG 2019   185
D. Munro, *DIY MEMS*, https://doi.org/10.1007/978-3-030-33073-6

Printed in the United States
By Bookmasters